THE · MACDONALD · ENCYCLOPEDIA · OF

THE · MACDONALD · ENCYCLOPEDIA · OF

Paolo Arduini
Giorgio Teruzzi

Macdonald

A Macdonald BOOK

© Arnoldo Mondadori Editore S.p.A., Milan 1986
© in the English translation
Arnoldo Mondadori Editore S.p.A., Milan 1986

Translated by Geoffrey Culverwell

First published in Great Britain in 1986
by Macdonald & Co (Publishers) Ltd
London & Sydney

A member of BPCC plc

All rights reserved
No part of this publication may be reproduced, stored in a retrieval system, or transmitted, in any form or by any means without the prior permission in writing of the publisher, nor be otherwise circulated in any form of binding or cover other than that in which it is published and without a similar condition including this condition being imposed on the subsequent purchaser.

British Library Cataloguing in Publication Data

Arduini, P.
The Macdonald encyclopedia of fossils.
I. Paleontology — Dictionaries
I. Title II. Teruzzi, G.
560'.3'21 QE703

ISBN 0-356-12567-X

Printed and bound in Italy
by Officine Grafiche A. Mondadori Editore,
Verona

Macdonald & Co (Publishers) Ltd
Greater London House
Hampstead Road
London NW1 7QX

CONTENTS

EXPLANATION OF SYMBOLS

Stratigraphic position

Each rectangle, identified by a letter, corresponds to a geological era A, Archeozoic (Precambrian); P, Paleozoic; M, Mesozoic; C, Cenozoic; Q, Quaternary). The color reveals the era in which the fossil lived

Habitat

continental

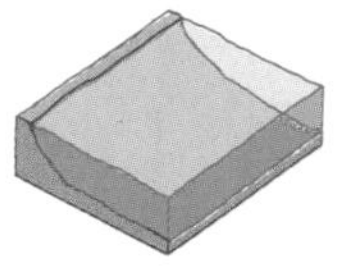

marine

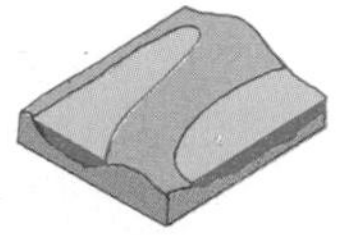

deltaic or lagoonal

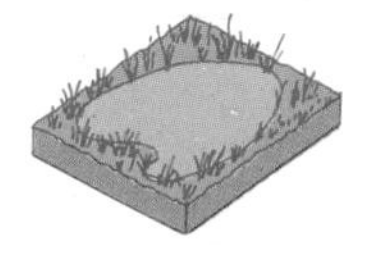

lacustrine or swamp

INFORMATION ON THE LAYOUT OF THE BOOK

This volume is divided into an introductory section, which provides general information necessary for tackling the subject, and an entries section. Each entry illustrates a group of fossils at the taxonomic level of the genus. This system provides the greatest possible amount of data on the greatest number of fossil groups (and thus the species included in each genus), which would not be feasible were individual species to be described. Working at the species level, in fact, would involve too much detail for a text that is basically intended to illustrate the main groups of organisms found in fossil form, with the sole exception of microfossils. The selection of genera is based principally on the need to provide an adequately representative résumé of what is known today about the organisms of the past. The genera chosen, therefore, are the ones most representative of individual animal and plant groups: they may be the commonest, the ones that best illustrate the basic characteristics of a particular group, or ones that display special peculiarities, either concerning the evolution of a specific group or general evolutionary phenomena.

The number of entries dedicated to individual groups is in more or less direct proportion to their importance. Sometimes, however, it has been deemed preferable to give space to organisms of particular interest (as, for example, in the case of jellyfish and, in the broadest sense, worms), which represent exceptional examples of fossilization (a factor that bears no relation to their rarity in fossil form, however important this may be from a paleontological point of view) and ones that are not normally dealt with.

The data contained within the entries has been arranged in the following way: a classification in accordance with contemporary systematics; a general description of the morphological characteristics of the genus, with particular reference, where appropriate, to the characteristics of the specimen illustrated; data on stratigraphic position and geographical distribution, with reference to the principal areas of discovery and the absolute age expressed in millions of years; and a note containing information of an ecological and evolutionary nature or other general remarks.

The illustrations and diagrams accompanying the individual entries act both as a complement to the entries with a reconstruction, where necessary, of the organism under examination, and also as a reminder of two other essential pieces of information: the age and environment of the genus.

The *Classification* section, which precedes the description of each genus, sometimes appears incomplete: this is because specialists have as yet not defined certain taxonomic entities above the level of genus.

INTRODUCTION

The term "fossil" is used to describe the remains of all animal or plant residues from the past, including any traces of their activities, which have survived up until the present day thanks to the physicochemical processes known as "fossilization." The branch of science concerned with the study of fossils is called "paleontology," a Greek word meaning literally "study of life," and involves the application of a number of disciplines associated with the natural sciences: zoology, botany, ecology, geology, biogeography, etc.

Fossils have been known to people for many centuries: fossil shells used for the purposes of adornment have been discovered at paleolithic sites, and during classical antiquity a number of writers, among them Strabo, Herodotus and Xenophanes, correctly identified vestiges of ancient organisms in petrified remains. However, the powerful influence exerted by the ideas of Aristotle and his school meant that, right up until the Middle Ages and beyond, fossils were on the whole regarded merely as freaks of nature formed spontaneously by some plastic force present in the primordial mud. Certain exceptionally enlightened minds disagreed with this school of thought: men such as Leonardo da Vinci (1452–1519), Gerolamo Fracastoro (1480–1533) and the Frenchman Bernard Palissy (1510–1590), all of whom recognized the true nature of fossils. The seventeenth century saw the appearance of the important writings of Agostino Scilla (1629–1700) and of Fabio Colonna (1567–1650), in which comparative studies were made of fossil forms and their living counterparts, and the work of the Dane Steno (1631–1686), who was the first man to formulate the law of successional sedimentary strata, whereby in a normal geological profile the lowest stratum is the oldest and the highest the most recent. Yet it was not until the eighteenth century that a correct interpretation of fossils became universally accepted: the publication of works laying the foundations of geology, on the one hand, and the inauguration of modern systematic zoology by Linnaeus (1707–1778), on the other, marked the birth of modern paleontology, to which the Frenchman Georges Cuvier (1769–1832) added a solid scientific basis with his studies on fossil vertebrates.

The process of fossilization

At the moment of an organism's death, the soft parts forming its body start a process of decomposition caused by the action of bacteria and by environmental circumstances leading ultimately to their total destruction; only rarely does this process fail to occur. The more durable parts of an organism (those containing a relatively high mineral content, such as bones, teeth, or shells) are able to last longer and to pass through those physicochemical stages that lead to their preservation in fossil form. As a rule, therefore, the elements with the greatest chance of becoming fossilized are the bones and teeth of vertebrates, the shells of molluscs and brachiopods, the exoskeletons of crustaceans and trilobites, the skeletons of echinoderms, the structures of corals and bryozoa, and the woody parts of plants. Clearly the surround-

Various examples of fossilization, from left to right and top to bottom: A silicified coral, a pyritized brachiopod, an ammonite with its original shell, a sea urchin in azurite, insects in amber and the (original) lower jawbone of an elephant.

This page: Carboniferous plant in siderite nodule (left) and a Mesozoic belemnite in vivianite (right). Opposite: Internal molds of gastropods.

ings in which the remains of a dead organism are deposited are important in determining the likelihood of the latter's survival as a fossil: generally, an underwater environment will be more favorable to fossilization than one on land.

The most widespread fossilization process is that of "mineralization": in this case the original substance is completely replaced by a mineral substance present in saturated form in the water permeating the sediment. The mineral substances most commonly participating in this process are calcium carbonate ($CaCO_3$)—for the most part in the form of calcite and, more rarely, aragonite—and silica, in the form of either quartz (SiO_2), opal, or chalcedony. Less common is transformation into pyrite or marcasite (FeS_2), into limonite (FeHO) or sulfates such as gypsum ($Ca[SO_4]2H_2O$), or into phosphates such as vivianite ($Fe_3[PO_4]_2-8H_2O$), or glauconite. A fossil preserved through a process of mineralization may take different forms:
if a shell becomes filled with sediment and then dissolves, we are left with an internal mold;
a pseudomorph is produced when the original shell is replaced by a different mineral substance;
an external mold or imprint is achieved when the shell leaves its exterior shape impressed on the sediment.

A process normally involving plant matter is that of carbonization, which occurs as a result of the action of bacteria in anaerobic conditions on large accumulations of organic substances. This leads to the elimination of oxygen and nitrogen and a

consequent increase in carbon and hydrogen. It is through this process that the great deposits of natural carbon (coal) were formed, most notably in the marshy forests of the Carboniferous period. Carbonization chemically alters the proteins and cellulose of tissues through degradation by bacteria, by chemical action, and by pressure and heat, until only carbon films remain. Other organic materials are dissipated as gases—carbon dioxide, methane, hydrogen sulfide, and water vapor. Coal is formed in the same way, but on a much larger scale.

Incrustation is a process of fossilization that occurs as a result of organisms steeped in calcium-rich water becoming coated in minerals such as travertine: when the organism itself dissolves, its imprint remains on the surrounding crust of mineral deposits.

More rarely, a process of distillation takes place, with the more volatile constituents of the organism becoming distilled, leaving behind a thin carbon film.

The instances of organisms being preserved *in toto*, complete with their soft parts, are very rare: when this happens, the result is that even the most delicate structures are preserved, like the bird's feathers discovered in the Jurassic limestone at Solnhofen in Bavaria. Creatures with no hard parts have been known to survive, such as the worms found in the Cambrian Burgess Shale formation in British Columbia, the Carboniferous layer at Mazon Creek, Illinois, and the Jurassic layer at Osteno in Lombardy. In a class all of their own are some types of total preservation, such as the mammoths trapped in the ice of Siberia or the insects impris-

Above: Arctica islandica*: a climatic indicator of cold climate.*
Below: Group of aligned fish, indicators of the existence of currents on the sea floor.

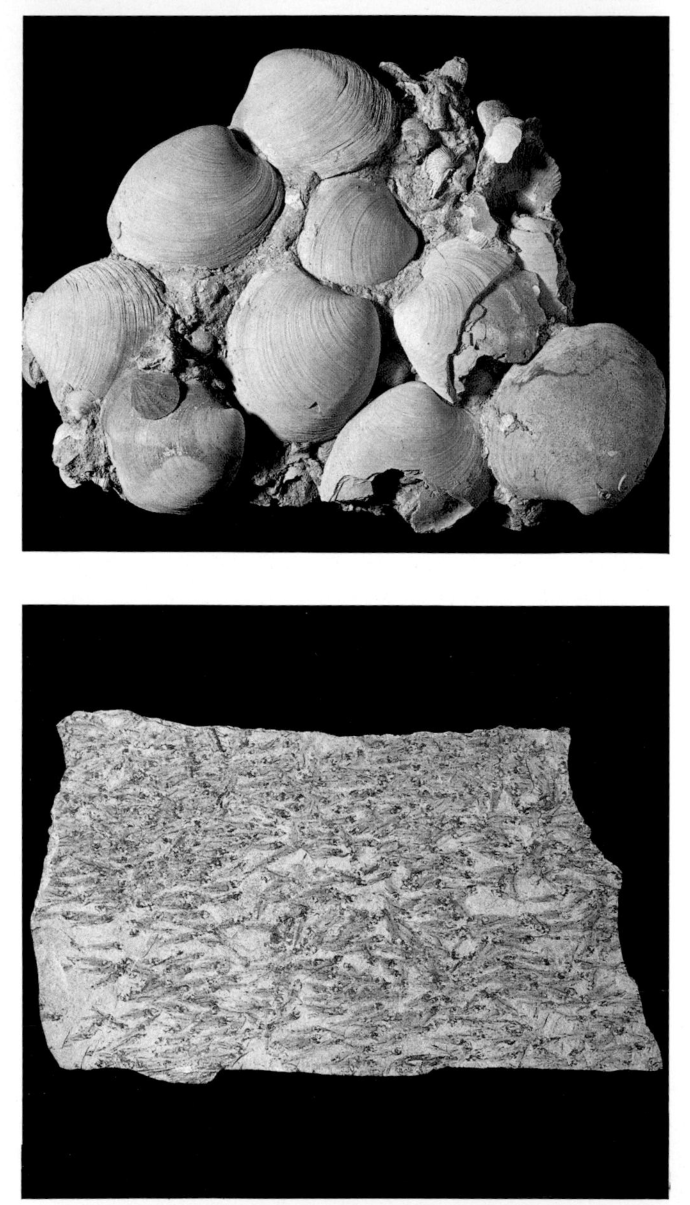

oned in amber (the ones found in Baltic amber are particularly famous). Natural mummification can also preserve parts of an animal that would otherwise perish: famous examples of this phenomenon are the dinosaurs of the genus *Trachodon* discovered in the Cretaceous of Wyoming, which had been first "mummified" and then mineralized.

Taphonomy

The discipline dealing specifically with the processes that occur from the moment of an animal's death up until the time that its remains become encased in sediment is known as "taphonomy." This science involves examination of all the data derived from a study of the way in which the fossilization processes have taken place in order to obtain ecological information. In the case of those shells or fish, for example, which display an alignment preference, information can be gleaned on the existence and nature of any possible currents. Conditions prevailing on a sea floor can be deduced from the presence or absence of tunnels and other traces left by benthonic and epibenthonic organisms, and from the existence or lack of minerals indicative of a specific chemical environment (the presence of pyrites, for example, indicates an anoxic bed). In addition, it is important to know whether the fossil has been deposited within its original habitat (in which case it is called "autochthonous") or whether it has been somehow displaced ("allochthonous").

Fossils and evolution

Fossils provide the most tangible proof of the evolution of living organisms. Although paleontological documentation of life in the past is far from complete, as indeed it always will be, fossils still demonstrate beyond any shadow of a doubt that living species are not fixed immutables, but are, on the contrary, the product of a very long series of changes whose history, through fossil remains, can now be reconstructed in outline and, in several cases, also in detail.

First and foremost, paleontological documentation allows us to trace the origins of at least some major taxa: among the vertebrates, for example, it has proved possible to document the overall relationship between reptiles and birds, reptiles and mammals, and between crossopterygian fish and primitive tetrapods. If we are familiar with the basic evolutionary stages of many animal and plant phyla, or at least those that have left extensive traces in rocks, it will also be possible to gain more detailed information on the evolutionary processes of lower taxa, such as species. By collecting different specimens of a single species, layer by layer, from the same fossil locality, it has in several cases proved possible to gain an overall view of the small evolutionary changes that have occurred within that species. Evolution at the taxa above species level is called "macroevolution," while "microevolution" is the name given to the evolutionary phenomenon observed at the level of the lowest taxa.

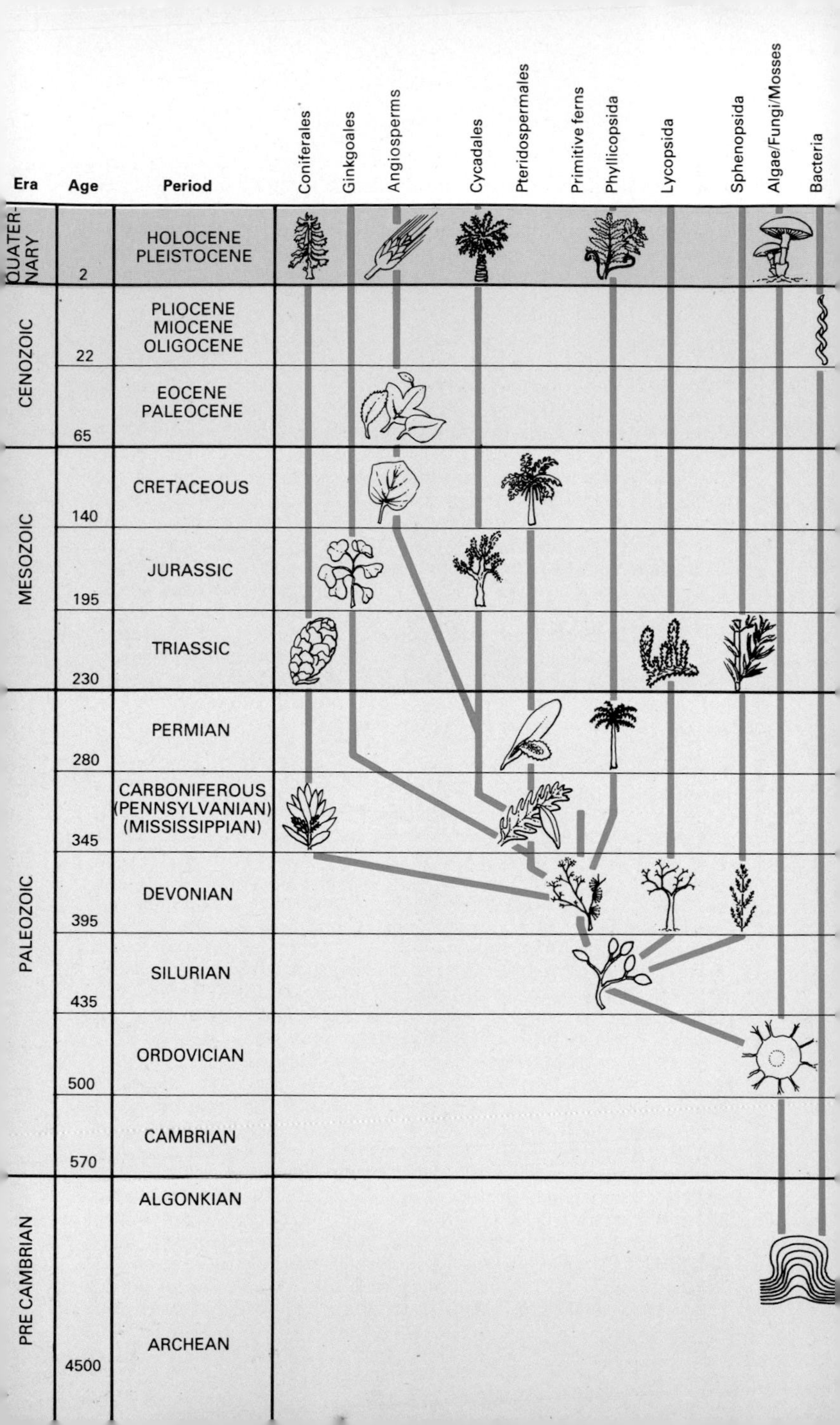

Era
Age
Period
Coniferales
Ginkgoales
Angiosperms
Cycadales
Pteridospermales
Primitive ferns
Phyllicopsida
Lycopsida
Sphenopsida
Algae/Fungi/Mosses
Bacteria
QUATER-NARY
2
HOLOCENE
PLEISTOCENE
CENOZOIC
22
PLIOCENE
MIOCENE
OLIGOCENE
65
EOCENE
PALEOCENE
MESOZOIC
140
CRETACEOUS
195
JURASSIC
230
TRIASSIC
PALEOZOIC
280
PERMIAN
345
CARBONIFEROUS
(PENNSYLVANIAN)
(MISSISSIPPIAN)
395
DEVONIAN
435
SILURIAN
500
ORDOVICIAN
570
CAMBRIAN
PRE CAMBRIAN
ALGONKIAN
4500
ARCHEAN

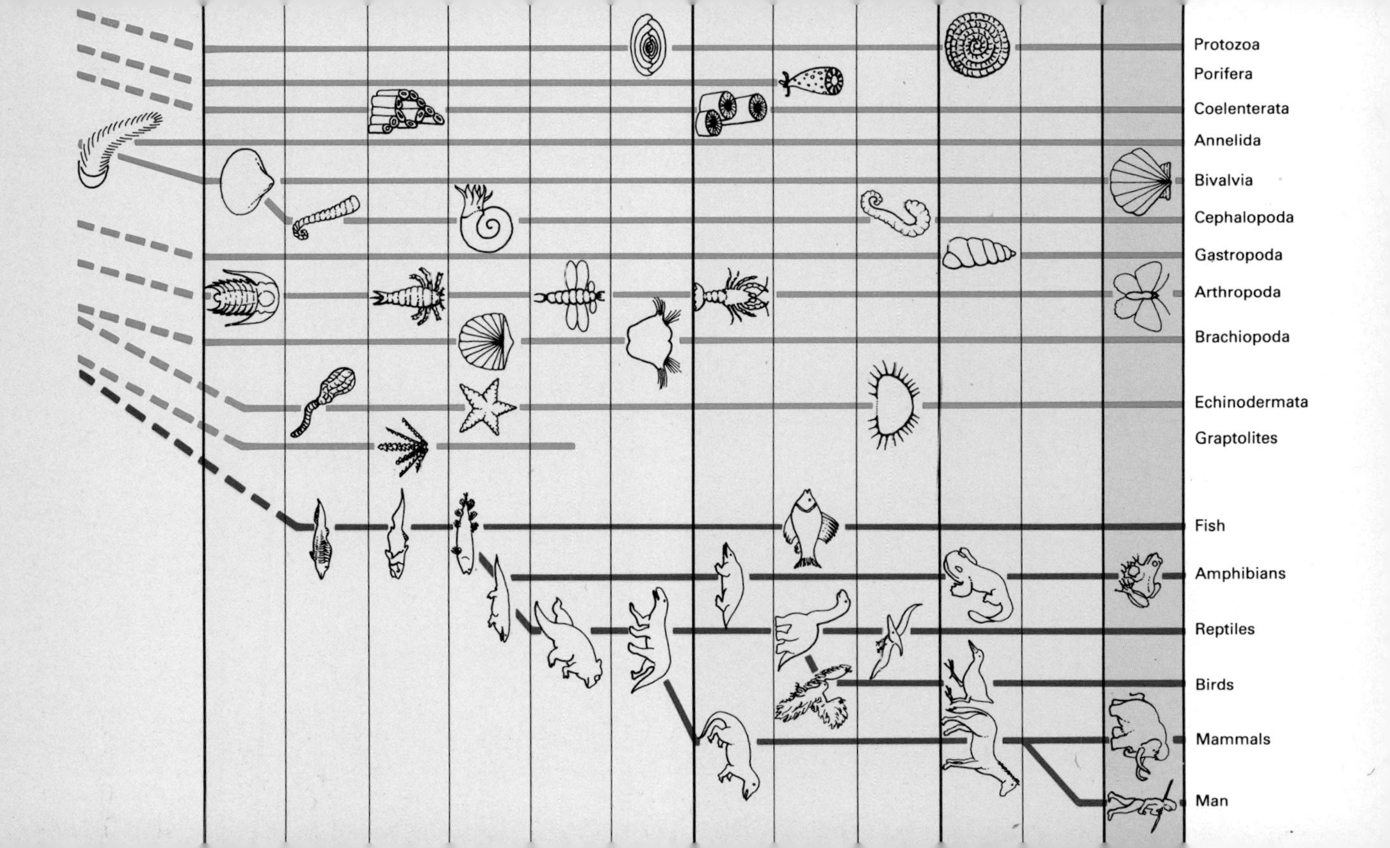
Protozoa
Porifera
Coelenterata
Annelida
Bivalvia
Cephalopoda
Gastropoda
Arthropoda
Brachiopoda
Echinodermata
Graptolites
Fish
Amphibians
Reptiles
Birds
Mammals
Man

Biostratigraphy

Biostratigraphy is concerned with the reconstruction of the exact succession of sedimentary rocks. This reconstruction is based, on the one hand, on the already-mentioned principle that in a normal succession of strata the oldest layer will be the lowest and the most recent one the highest. The word "normal" here refers to an ideal situation in which successive deposits have not been subjected to tectonic disturbances, where, for example, there has been no alteration to the order in which the strata were originally laid down. Apart from this geometrical principle, the successional reconstruction of sedimentary rocks is also based on differences in paleontological content resulting from the gradual evolution of living organisms. In the earliest rocks primitive organisms strikingly different from modern forms become more and more similar in appearance to modern-day flora and fauna toward the present.

In this way it has proved possible to reconstruct the sequence of physical and biological events that have characterized the history of the earth and to subdivide it into smaller units called "periods," "epochs," and "zones;" the latter two further subdivided into "subepochs," "subzones," etc. In this way a relative age can be ascribed to every event in the earth's history. The last of the dinosaurs, for example, became extinct in the Cretaceous period, at the end of the Maastrichtian epoch, which marks the close of the Mesozoic era.

In order to arrive at an absolute date, expressed in millions of years, for example, paleontologists resort to radiometric methods, which measure the decay of radioactive elements. By means of these methods, which involve the use of uranium, potassium and strontium isotopes, it is generally possible to find out the age of igneous rocks and, in special cases, that of sedimentary rocks in which fossils are usually found. By combining the two types of chronology, absolute and relative, a geological timetable has been obtained of the type illustrated on pages 16–17.

The fundamental unit of biostratigraphy, the zone, is defined as a group of rock layers characterized by the presence of a single fossil species or a characteristic association of species. A zone, therefore, represents the period of time during which a species or an association of species existed. Not all types of fossil can be used for this method of dating, however. Fossils with a restricted geographical distribution or an excessively long period of existence are on the whole no use in establishing a biostratigraphical subdivision for a fairly large geographical area. On the other hand, fossils with a broad geographical distribution and of short chronological duration are ideal for the purposes of biostratigraphical subdivisions and are known as "guide fossils." Various different groups of fossil organisms are used for these biostratigraphical subdivisions, with the times for which they are used coinciding with their most successful periods (i.e. those times when a group is represented by numerous members with species that evolve at a very rapid rate), thereby making it easier to create subdivisions and correlations on a global scale. Thus, in early Paleozoic subdivisions trilobites and graptolites are extensively used; in the Mesozoic ammonites are used, and in the Cenozoic and Quaternary, Bivalvia. From the Cretaceous onward, microfossils begin to assume a very important role. Vertebrates, on the other hand, are rarely used, since they are often preserved only in very exceptional cases.

Fossils and paleoecology

Based on the supposition that natural phenomena occurred in the past in the same way as they occur now, in accordance with what is known as the "uniformitarian" or "actualist" hypothesis, it is possible, on the basis of a study of the characteristics of the fossils contained within it, to reconstruct the original environment in which a sequence of sediment was laid down, combined, of course, with a study of its geochemical and sedimentological characteristics.

These depositional environments can be subdivided into marine and continental, with the further addition of special transitional environments, such as deltas and lagoons. Continental environments can be lacustrine, fluvial, glacial, desert, steppe, savanna, cave, etc., while their marine counterparts can be subdivided on the basis of water depth into littoral, neritic (up to 200 meters or 110 fathoms below sea level), bathyal (up to 2,000 meters or 1,100 fathoms) and abyssal (below 2,000 meters or 1,100 fathoms).

The information that fossils are able to provide relates to all the characteristic components of an environment: degree of temperature, salinity, etc. Associations of fossilized terrestrial flora, for example, generally furnish accurate information on climatic conditions, while the alternation in Quaternary deposits between the remains of vertebrates adapted to a warm climate and those adapted to a cold one provides classic evidence of the climatic changes that occurred during the Quaternary period.

One excellent indicator of climate are the corals and other organisms that contribute to the building of coral reefs. At the present time, these reefs develop solely in zones characterized by specific environmental conditions: temperature between 25°C (77°F) and 29°C (84.2°F), salinity varying between 34 and 36 parts per thousand in a depth of not more than 50 meters (27 fathoms). It is believed that in the past fossil reefs formed under similar conditions.

A special marine environment is represented by the euxinic type, which takes the form of an inclosed basin (like the modern Black Sea), in which, because the water is never renewed, there is a lack of free oxygen and an anoxic environment is created. Under these specialized conditions there is a complete absence of benthonic organisms, and the remains of the animals living in the surface waters of the basin, the only place in which it would be possible to survive, are deposited on the sea floor, where they lie undisturbed owing to the lack of any necrophagous organisms, a phenomenon that favors the likelihood of preservation.

These basins also offer favorable conditions for the formation of iron sulfides, such as pyrite, a mineral that often plays a part in fossilization in these environments.

Marine benthonic organisms, meaning those whose lives are linked to the sea floor, and in particular molluscs, because of their abundance, provide information both on the environment in which they were laid down and also on the actual depth of the sea, since, as we have already seen in the case of corals, they live within well-defined depths.

Fossils and paleogeography

Paleogeography deals with the reconstruction of the geographical landscapes of the past. This relates primarily to the changes that have occurred in the seas and land masses during the course of the different geological periods, but paleogeography is also concerned with the reconstruction of the past distribution of flora and fauna, deserts, mountains, lakes, climates and all the other factors that contribute to establishing the characteristics of a geographical landscape at a given moment in the earth's history.

One of the great causes of geographical change in our planet seems to have been the birth of the continents: according to the "plate tectonics" theory, the earth's lithosphere is divided into moving sections that include the continents, with the result that the continents' positions vary with the passing of geological time. Fossils can provide classic evidence of this continental drift, through, for example, the discovery of remains of the same organisms, either terrestrial or at least incapable of traveling for long distances across the sea, in continents today separated by oceans. Such is the case with fossils of *Lystrosaurus*, a typically terrestrial mammal-like reptile of the Permian period, whose remains occur in Antarctica, India, South America and southern Africa, thereby indicating that these land masses were joined together at the end of the Paleozoic era. Another classic example is provided by the mesosaurs, small reptiles discovered in Permian rocks in Brazil and southern Africa: further evidence that southern Africa and South America may have been linked together at that time.

Where fossils are found

Fossils are generally found in sedimentary rocks, those rocks formed by an accumulation of debris collected from the breakdown of other pre-existing rocks (shale, sandstone, conglomerates) or by the chemical precipitation of minerals, as in carbonate rocks. It is, nevertheless, not uncommon to discover fossils in certain igneous rocks such as tufa, or in rocks that have undergone a certain degree of metamorphism.

Fossils may be found completely by chance, although it is true that there are certain places with a greater potential: where natural erosion has uncovered the rock in the sides of cliffs or valleys, for example, or where exposures have been created along roads and highways.

Manmade excavations provide a very fertile hunting ground,

An example of sedimentary strata: The Grand Canyon in the state of Colorado.

whether these take the form of quarries or mines, or whether they are the result of road building, construction work or tunneling.

Fossils are extracted by different means, depending both on the type of fossil involved and on the nature of the rock encasing it. Great care should always be taken in the removal of a rock-encased fossil: a wide surrounding area should be excavated in order to extract it intact, together with a section of the matrix, while cleaning should be carried out in a specially equipped laboratory. The implements used for extraction from hard rocks are mallets, hammers (the typical paleontologist's hammer has a flat head) and chisels of different shapes and sizes. Other equipment needed are a magnifying glass, brushes, adhesive, specimen bags, paper in which to wrap the finds and labels on which to write the location and layer of the fossil's discovery: for more extensive annotations a proper notebook is also required. Permission should be obtained before collecting on private property. Several areas have regulations or laws which forbid or restrict collecting. Persons interested in collecting should familiarize themselves with these restrictions.

The preparation of specimens

Fossils are generally discovered partially imbedded in the surrounding rock. In this case it is necessary to carry out the process known as "preparation," which is performed using the methods appropriate to the type of rock involved.

For some fossils this preparation is a very simple matter: in the

Digging for fossils.

case of those imbedded in sand, for example, a quick rinse is often enough to get them clean. In the majority of cases, however, especially where the encasing rock is calcareous, preparation is a long, difficult and painstaking process. These conditions call for the use of hammers, some of them very small and light, and a variety of scalpels of different shapes and sizes. A vibrating electric stylus may also prove useful to speed up the work. After removal of most of the inclosing matrix comes the fining-down, which has to be very accurately carried out in order to avoid irreparably damaging the specimen. The ideal tools for this stage are strong, pointed implements, such as the needle of a record player or an entomologist's pin, and thin awls and small cutting blades in order to remove the final layer of matrix. This final process is performed using a low-power binocular microscope (5×–30×). During preparation it is often necessary to use powerful, transparent glues of the type used for making models.

Top to bottom: Various stages in the exposure of a fossil reptile.

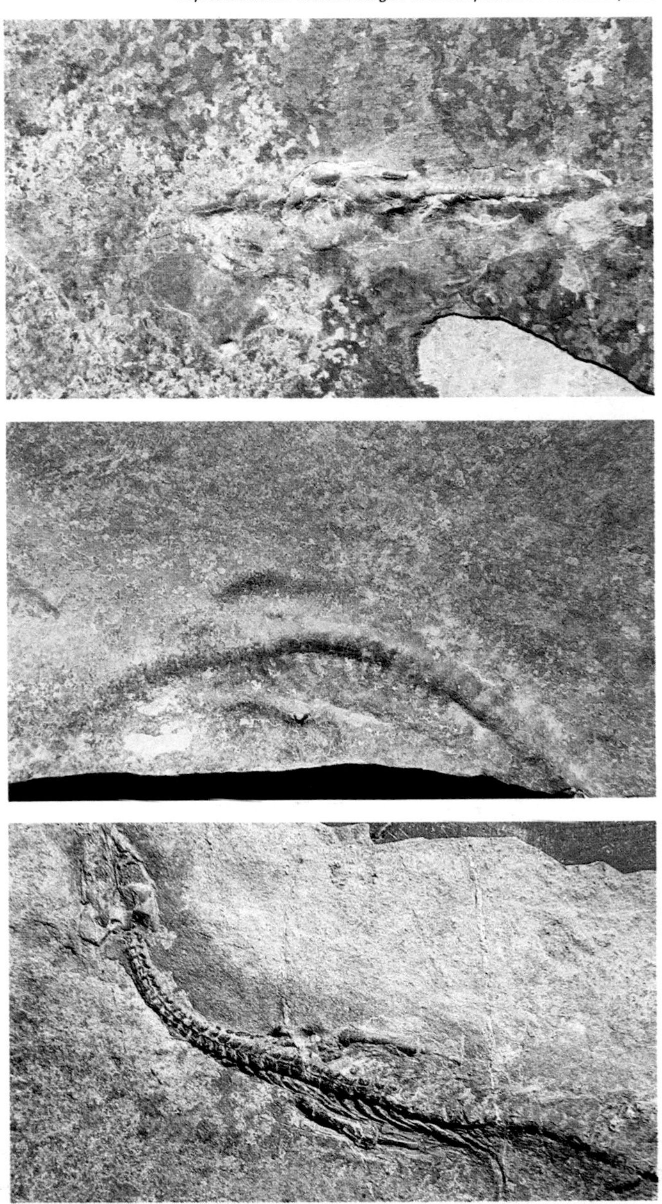

CLASSIFICATION

The system of nomenclature used today to define living and fossil species is the one introduced by the Swedish naturalist Linnaeus: it involves designating each species with two names and is therefore called "binomial." Thus *Hildoceras bifrons*, for example, is the scientific name of a certain species of ammonite: it is composed of the generic name *Hildoceras*, normally written with a capital letter, and the specific name *bifrons*, written in lower case.

The animal kingdom is subdivided into large systematic categories called *phyla* (singular *phylum*), which can be further subdivided into classes, orders, families, genera and, finally, species. The species, the fundamental unit of systematic classification and the only natural one, is defined in the case of living organisms on the basis of the interfertility of its individual members. In the case of paleontological species, for which this criterion does not apply, a species is normally defined by statistical measurement of the variability of its constituent members. When this is impossible because of inadequate number of individual specimens, emphasis is placed on those analogous characteristics that indicate certain individuals may belong to the same species. The latter criterion will inevitably reflect the subjectivity of the individual classifier, since one paleontologist may consider important a characteristic that another may regard as irrelevant; on the other hand, a similar imperfection in classificatory criteria is implicit in the type of material to be classified.

A living or fossil species is given a name by its discoverer: this word is then Latinized. A generic or specific name can designate certain peculiarities of the organism in question, such as *Micraster* ("a small star"), or it can reflect the area in which it was discovered, such as *Nigericeras* or *Texanites*, or it can even be dedicated to a particular scholar, such as *Sowerbyceras* or *Oppelia*.

If systematic categories need further subdivisions, it becomes necessary to use subgenera, subspecies, subclasses, etc. In paleontology there is also frequent use of qualifying terms such as *Lebachia cf. piniformis* or *Hippurites aff. taburnii*, in which *"cf."* and *"aff."* indicate a similarity or affinity with another species, at the same time underlining the fact that the organisms in question do not possess characteristics that coincide exactly with those of the species with which they have been bracketed.

PLANT KINGDOM

The list below provides a brief summary of the most significant plant groups from a paleontological point of view.

Phylum Cyanophyta

Includes blue-green algae, organisms made up of several cells forming filamentary aggregates. They give rise to calcareous structures in the form of concentric incrustations, called Stromatolites. Known since Precambrian times.

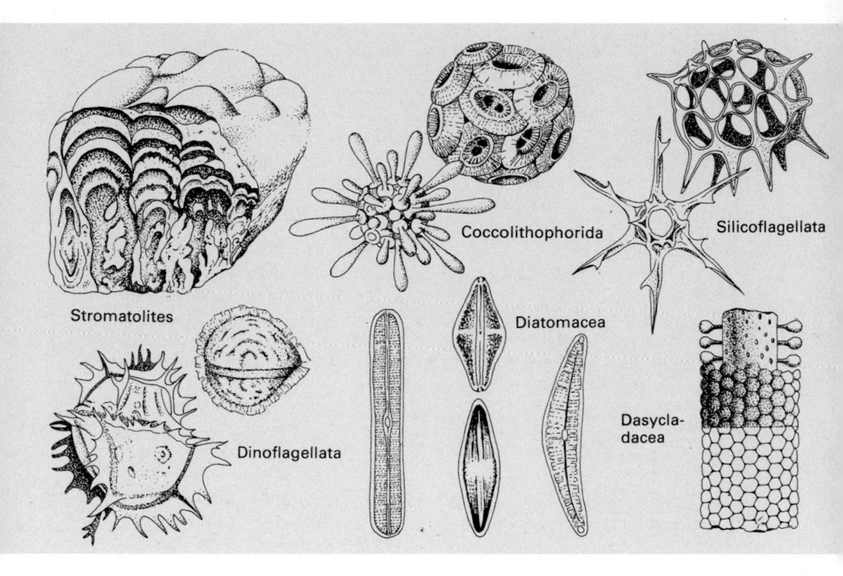

Phylum Schizomiceta
Includes bacteria, simple organisms whose actions are responsible for the formation of certain deposits of iron, limestone and phosphates. Known since Precambrian times.

Phylum Chrysophyta
Includes organisms that are relatively common in fossil form: Coccolithophorids and Silicoflagellates. Coccolithophorids, now found mostly in warm seas, consist of flagellate cells covered by calcareous disks and have been known since Paleozoic times. Silicoflagellata, by contrast, which have a siliceous skeleton, have been known since Cretaceous times.

Phylum Pyrrophyta
Includes the Dinoflagellates, planktonic organisms possessing a theca with a variety of ornament. Known since Paleozoic times.

Phylum Bacillariophyta
Includes the Diatomaceans, microscopic organisms whose siliceous shell forms white, friable deposits called "tripoli" or "rotten-stone" (or diatomaceous earth). Known since Jurassic times.

Phylum Rhodophyta
Comprises red algae, including the Solenoporacea, widespread from the Silurian to the Cretaceous, and the Corallinacea, known since

Cretaceous times up to the present. They were important in the past as builders of underwater reefs.

Phylum Chlorophyta
Includes the green algae, now widespread in tropical seas and fresh water. They were very important in the past as builders of limestone reefs, particularly in the form of the groups Codiaceae and Dasycladaceae.

Phylum Tracheophyta
Includes the vascular plants, subdivided into numerous classes, among them:
Class Psilopsida. These were the first land plants, which appeared in the Silurian era and became very widespread in the Devonian. They had a subterranean stalk, bereft of roots, which formed vertical and dichotomously divided aerial stalks.
Class Lycopsida. Represented today by a few herbaceous forms, this class contains treelike forms that during the Carboniferous era reached heights of 30 meters (97½ ft), such as *Lepidodendron* and *Sigillaria*.
Class Sphenopsida. Represented today by Equisetaceae, small herbaceous plants, they developed treelike forms during the Paleozoic era that became widespread during the Carboniferous era.
Class Phyllicopsida. Fernlike plants that were very important during the Carboniferous period and are still widespread today.
Class Gymnospermopsida. Woody plants that appeared during the Devonian period and which do not reproduce by means of spores, but possess seeds with no protective covering. This class contains different orders:
Order Pteridospermales. Plants similar to ferns, but possessing seeds, which became totally extinct at the end of the Paleozoic era.
Order Cycadeoidales. These first appeared during the Carboniferous period and are still widespread in the tropics. Similar to palm trees, they were abundant in the Triassic and Jurassic periods.
Order Ginkgoales. Forest trees with a characteristic lobate leaf, the sole surviving representatives being *Ginkgo biloba*.
Order Cordaitales. Primitive gymnosperms that appeared during the Devonian period and became extinct at the end of the Paleozoic. They reached considerable heights (30–40 meters or 98–130 ft) and possessed a trunk with branches bearing spirally arranged, elongate leaves.
Order Voltziales. Similar to modern-day Araucarias, this order appeared during the Carboniferous era and became extinct at the beginning of the Jurassic.
Order Coniferales. Very widespread today, these plants appeared in the Triassic era and achieved their greatest diversification during the Mesozoic.
Class Angiospermopsida. Representing the most advanced evolutionary stage in the plant world, these bear flowers and seeds contained within pods; they are also the commonest forms of vegetation on earth. The class contains two subclasses:
Subclass Dicotyledones. Plants bearing two seed-leaves (coty-

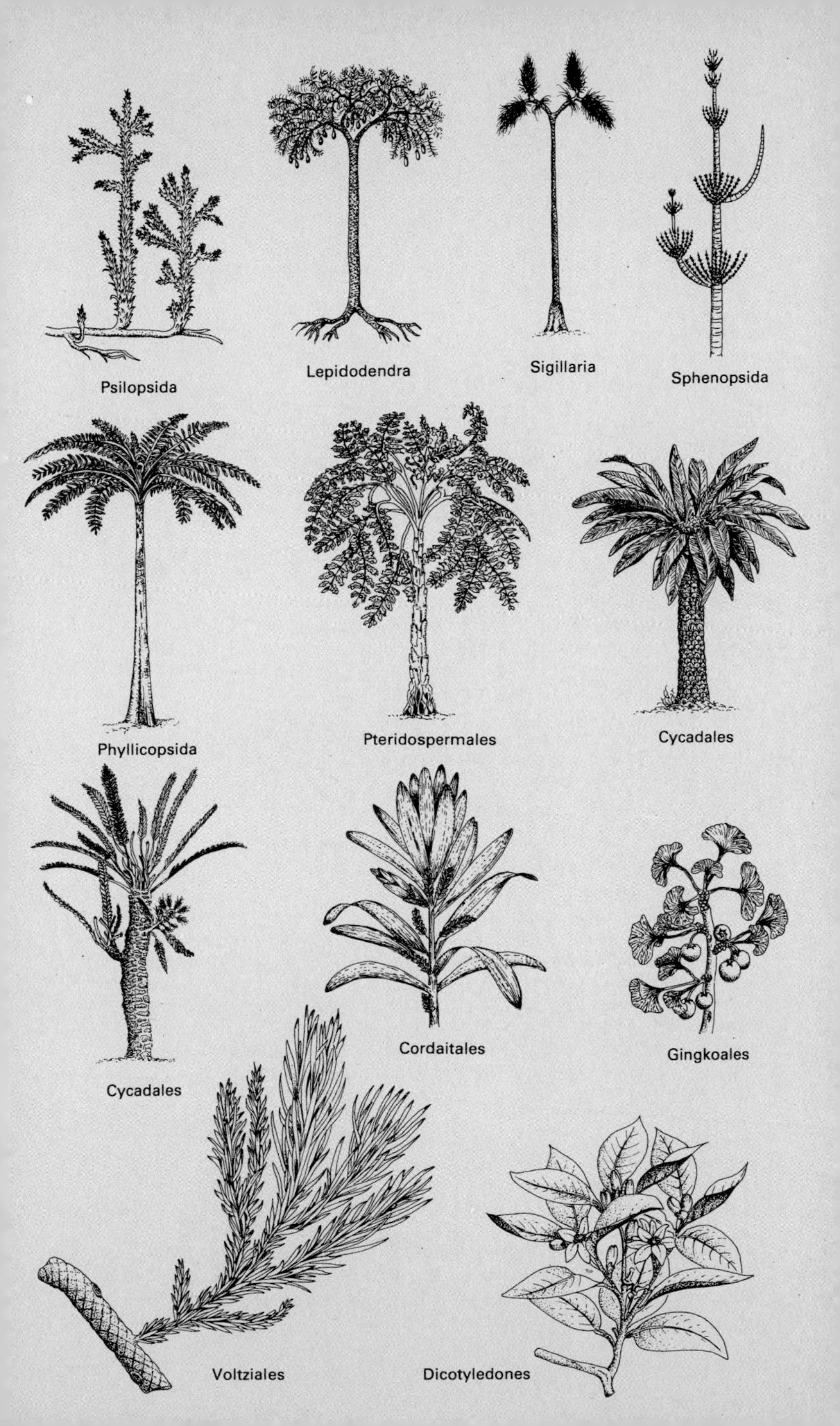

Psilopsida
Lepidodendra
Sigillaria
Sphenopsida
Phyllicopsida
Pteridospermales
Cycadales
Cycadales
Cordaitales
Gingkoales
Voltziales
Dicotyledones

ledons), which have been definitely identified from the Cretaceous period, although there are also doubtful remains dating back to the Jurassic.
Subclass Monocotyledones. Plants with one seed-leaf, known since the Cretaceous period.

ANIMAL KINGDOM

Phylum Protozoa
Protozoa are single-celled organisms whose dimensions range from a micron to several centimeters. The only known fossil forms are those which possess a siliceous, calcareous or chitinous shell. They are subdivided into numerous classes:
Class Foraminifera. These are without doubt the most important protozoans from a paleontological point of view. They are marine organisms, usually with a calcareous shell, which have provided a large number of guide fossils since the Cambrian era.
Class Radiolaria. These possess a siliceous (opaline silica) shell of variable shape, characterized by spines and processes. These protozoans first appeared during the Cambrian era and are now found in all the seas of the world.
Class Ciliata. Protozoa with a bell- or cup-shaped shell whose chemical composition is thought to be pseudo-chitinous. They are known in the fossil state solely through the tintinnid group, examples of which are very plentiful in rocks of the Upper Jurassic and Lower Cretaceous periods.

Phylum Porifera
Porifera—better known as sponges—are marine animals, either single or grouped together in colonies, multicellular and asymmetric. They possess no nervous system, no internal organs, no mouth and no anus. They have a calcareous or siliceous skeleton made up of spicules that fuse together to form a rigid frame. The walls of this frame possess numerous inlets that communicate, either directly or through a series of canals, with the gastric cavity, into which food is drawn by means of flagellae, which cause a swirling current of water. In the adult stage all sponges live either attached to the sea floor or affixed to marine creatures. Porifera are known since the Precambrian era.

Phylum Archeocyathida
Archeocyathida are an extinct group of organisms that inhabited the seas of the Cambrian period. They possessed a conical or discoidal skeleton made of calcite and were able to congregate in colonies fixed firmly to the sea floor. The skeletal structure of Archeocyathida is similar to that of sponges in its general shape and to that of Coelenterata in the occurrence of radial plates, called "septa," which run the length and breadth of the test, the walls of which are porous.

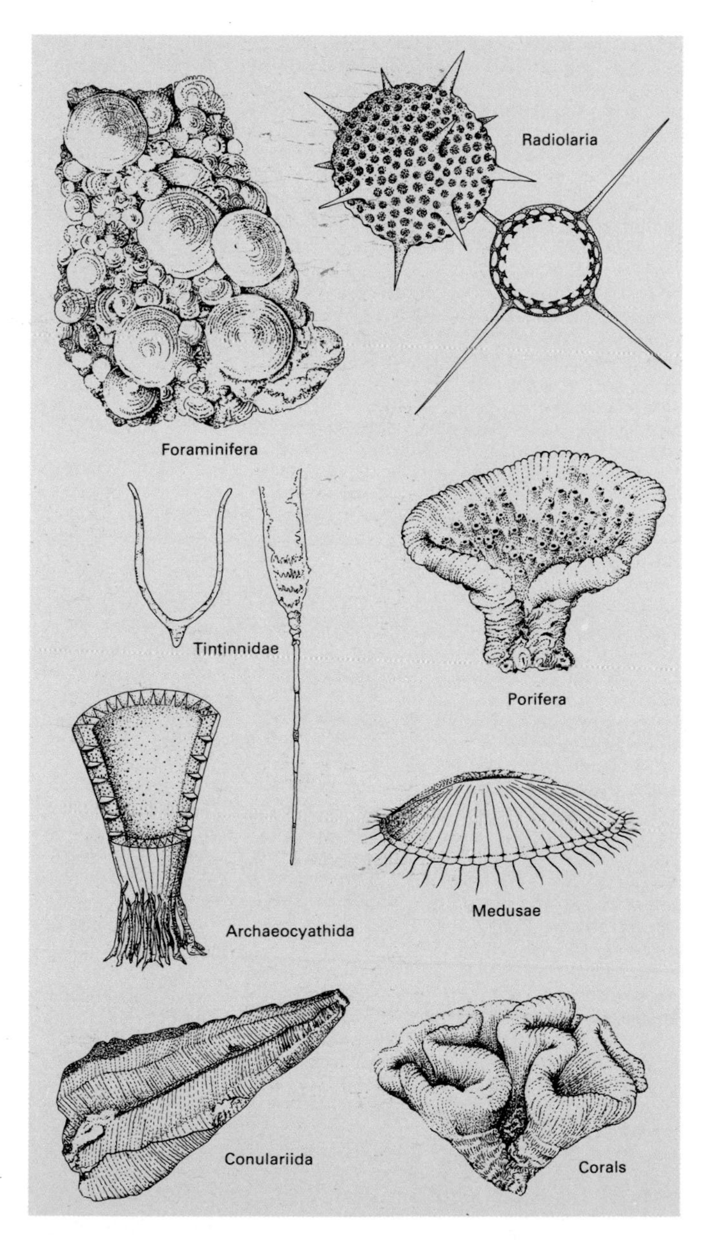
Radiolaria
Foraminifera
Tintinnidae
Porifera
Archaeocyathida
Medusae
Conulariida
Corals

Phylum Coelenterata

The coelenterates include jellyfish and sea anemones, which are soft-bodied, and corals, which possess a calcareous skeleton. Their bodies are composed of two layers of tissue inclosing a central cavity, the *coelenteron*, and they are equipped with a mouth surrounded by retractile tentacles. They are divided into three classes: Scyphozoa (jellyfish), Hydrozoa (sea anemones) and Anthozoa (corals). The corals are the most interesting from a paleontological point of view and comprise: Rugosa, Paleozoic corals with a calcareous skeleton containing four primary septa and other, secondary ones which develop during maturity; Scleractinia, corals with a calcareous skeleton in which the septa develop in cycles of six and successive multiples of six; Tabulata, which appeared in the Ordovician period and vanished at the end of the Paleozoic, whose individual corallites are partitioned by transverse tabulae.

Phylum Bryozoa

The bryozoans are aquatic, sessile animals that form colonies of various shapes both in sea and fresh water. Very widespread in the modern day, their fossils are known from the Ordovician period. They are similar to the coelenterates, but possess a much more complex body structure, with a complete alimentary canal, mouth and anus. Their skeleton is calcareous and is secreted by individual members assembled together in a colony.

Phylum Brachiopoda

Brachiopods are filtering invertebrates found solely in salt water; they possess a bivalved shell and live on the sea floor, to which they either affix themselves by means of a pedicle emerging from one of the valves or, more simply, they just rest on it. The shell, which completely covers all their soft parts, may be either calcareous or chitinophosphatic. Brachiopods possess a very complex internal anatomy: within the valves there is a system of ciliate tentacles whose movement allows the animal to draw water into the shell, where filtration and respiration take place. Brachiopods are divided into two classes on the basis of the presence or absence of a hinge linking the two valves: the inarticulate brachiopods, which lack this hinge, are the more primitive; and in these the reciprocal movement of the two valves is achieved by means of a complex muscular system. The articulate brachiopods, on the other hand, have a hinge very similar to that of certain Bivalvia.

The earliest brachiopods, discovered in sedimentary strata of the Lower Cambrian era, belong to the inarticulate class; the articulate ones appear later in the Upper Cambrian. The greatest period of brachiopod development was the Paleozoic era, at the end of which they underwent a severe biological crisis. They survive today in a very small number of genera.

Phylum Annelida

Wormlike, segmented organisms. From a paleontological point of view, the most important annelids are the polychaetes. The soft-bodied, "wandering" polychaetes are known rarely as complete fossil

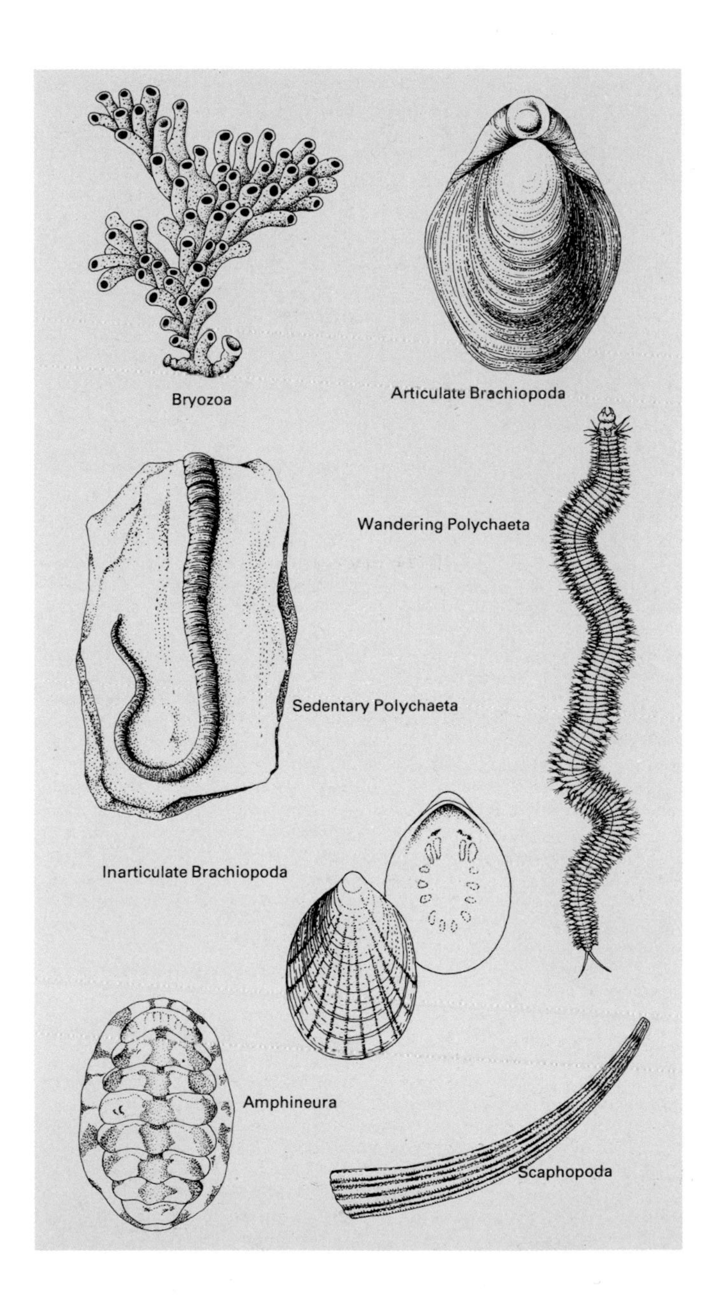
Bryozoa
Articulate Brachiopoda
Wandering Polychaeta
Sedentary Polychaeta
Inarticulate Brachiopoda
Amphineura
Scaphopoda

specimens or, more commonly, as fossilized jaws found in sedimentary strata (scoledonts). "Sedentary" polychaetes are known from the remains of the burrows and tunnels that they dug in sedimentary deposits. Polychaetes are known from the Precambrian period onwards.

Phylum Mollusca

Highly important, both because of the ease with which their calcareous shells are preserved and also because of the paleoecological and biostratigraphical data they provide. Molluscs are subdivided into six classes:

Class Monoplacophora. Monoplacophores possess a shell composed of a single, conical valve with bilateral symmetry. They appeared in the Lower Cambrian era and still exist today. They are the most primitive form of mollusc.

Class Polyplacophora. Polyplacophores or "chitons" are molluscs distinguished by their possession of a shell composed of seven or eight articulated calcareous plates. They first appeared in the Cambrian era and are still found today.

Class Scaphopoda. Scaphopods are a very small group of molluscs that first appeared in the Lower Devonian era and still occur in seas today. They are organisms with a single-valved shell, tubular in shape, slightly curving and open at both ends. Scaphopods lack eyes, but possess a ciliate proboscis. They are an intermediary group between bivalves and gastropods.

Class Bivalvia. Bivalves or Lamellibranchia are aquatic molluscs, essentially marine, possessing an outer shell, formed of calcium carbonate and conchiolin, which is composed of two valves. These valves articulate by means of a hinge and are joined together by a ligament. One or two adductor muscles work on the valves from within: their action counteracts that of the ligament, which tends to open the valves. A layer of tissue, the mantle, secretes the shell and incloses the soft parts. Bivalves have no head distinguishable from the visceral mass and are equipped with a foot used for digging and moving. Bivalves have different lifestyles: some are capable of swimming by agitating their valves, some attach themselves by a byssus to floating objects, while others cement a valve to rocks. Yet others live partially or completely buried in sediment. The taxonomy of bivalves is based on the form of the hinge. The earliest representatives of this class appeared in the Cambrian era and became very widespread in successive eras.

Class Gastropoda. Gastropods are molluscs adapted to marine, freshwater and terrestrial life. They have a head, distinct from the body and equipped with eyes, and they crawl forward by contracting their single foot. The shell, which is not common to all gastropods, has a single valve and is composed of calcium carbonate; it may coil in the form of a trochoidal spiral, as in the majority of cases, and is not divided into chambers. The classification of gastropods is not based on the characteristics of the shell, which can vary widely within a single genus, but on the soft parts; for this reason paleontologists use modern-day examples to subdivide the large fossil groups. Gastropods first appeared during the Lower Cambrian.

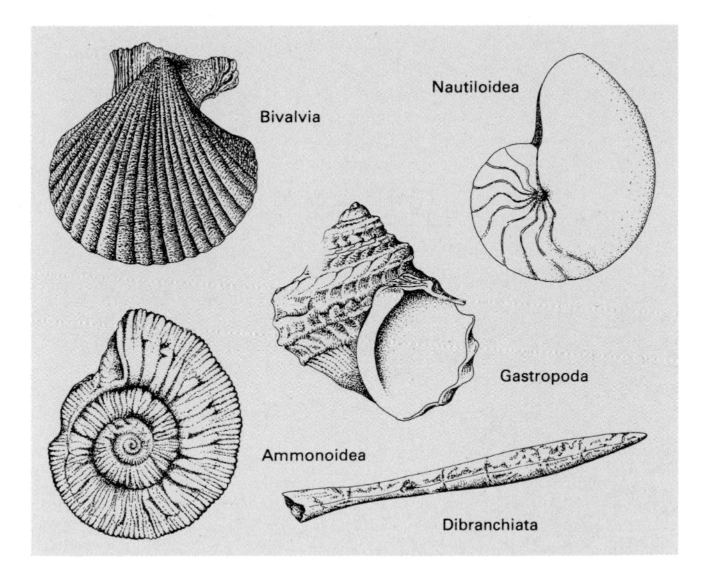

Class Cephalopoda. Cephalopods are exclusively marine molluscs, characterized by bilateral symmetry, a head clearly distinguishable from the body and a mouth region surrounded by tentacles. The head is equipped with two large, laterally placed eyes, while the mouth possesses jaws and radulae, both of which are suited to mastication. By rhythmically ejecting water from the mantle cavity from the hyponome, a structure placed beneath the mouth, cephalopods move in what appears to be a backward direction. Respiration occurs through gills, the number of which is itself a systematic characteristic: Tetrabranchiata or Ectocochlia (external shell), which have four gills; Dibranchiata or Endocochlia (internal shell), which have only two. Nautiloids and ammonites belong to Tetrabranchiata, whereas cuttlefish and squid belong to Dibranchiata.
Subclass Nautiloidea. The first nautiloids appeared during the Lower Ordovician era and are still found in seas in the form of the genus *Nautilus*. Their shell is external and subdivided into chambers by septa that display a very simple configuration; the outer surface is in the shape of an orthocone or "nautilocone" and has little ornament.
Subclass Ammonoidea. The first ammonoids appeared during the Lower Devonian and became extinct in the Upper Cretaceous. Their shell is external, generally planospiral in form; the septa have an undulating surface, which in the course of evolution became increasingly more complex in shape. The systematic classification within this group is based on the complication of the septa, the position of the siphuncle and the shape, configuration and ornament of the shell.

Subclass Dibranchiata. This subclass contains those cephalopods, either with or without an internal shell, which have developed since the Carboniferous period; its modern representatives are cuttlefish, squid and octopuses. A very important group in the past was the belemnites, creatures bearing a strong resemblance to squid, which possessed a very strong inner shell, and it is this shell which has endured most readily in fossil form.

Phylum Arthropoda

Arthropods are segmented animals with a chitinous external skeleton and paired, jointed limbs. In numbers of species it is the largest phylum in the animal kingdom, embracing insects, spiders, crustaceans, millipedes and the extinct trilobites. Arthropods possess bilateral symmetry and a body composed of articulated segments that each carry a pair of appendages that can fulfill a variety of functions. Arthropods inhabit every type of environment—aquatic, aerial, terrestrial—and date from the Cambrian period. They include the following classes:

Class Arachnida. Spiders and scorpions, mostly in terrestrial forms, which date from the Silurian era.

Class Merostomata. Represented in modern times by a single marine genus, *Limulus*, which belongs to the Xiphosurid group; includes the extinct group of Eurypterids, which were aquatic forms abundant during the Silurian and Devonian periods, some of whom developed into gigantic creatures as much as 2 meters (6.5 ft) long.

Class Insecta. The largest group of living arthropods and the only ones capable of flight. They are known in fossil form from the Devonian and are often found preserved *in toto* in Oligocene ambers.

Class Crustacea. Terrestrial and aquatic arthropods, this class comprises decapods (shrimps and crabs), which date from the Carboniferous period, and a number of other groups, including ostracodes, which are of great stratigraphical importance.

Class Trilobita. Totally extinct Paleozoic organisms representing the most important arthropod group for the paleontologist. They had a body divided longitudinally into three lobes: one central and two lateral. There were three recognizable sections (from front to back): the cephalon, the thorax and the pygidium. They were essentially marine and generally small in size (5–8 cm or 2–3 in on average), although some forms did reach 70 cm (27.5 in). First appearing in the Cambrian era, they reached their acme in the Ordovician, only to go into a slow decline in the Silurian, leading ultimately to their extinction in the Permian era.

Phylum Echinodermata

Echinoderms are exclusively marine animals, benthonic in a great many cases, either fixed or free-living. They possess pentamerous or bilateral symmetry; the exoskeleton is composed of contiguous plates, rarely imbricated, made of calcium carbonate, with external ornament provided by spines and tubercles. They are characterized internally by a water-bearing system made up of canals that communicate with the exterior of the animal through a network of pores.

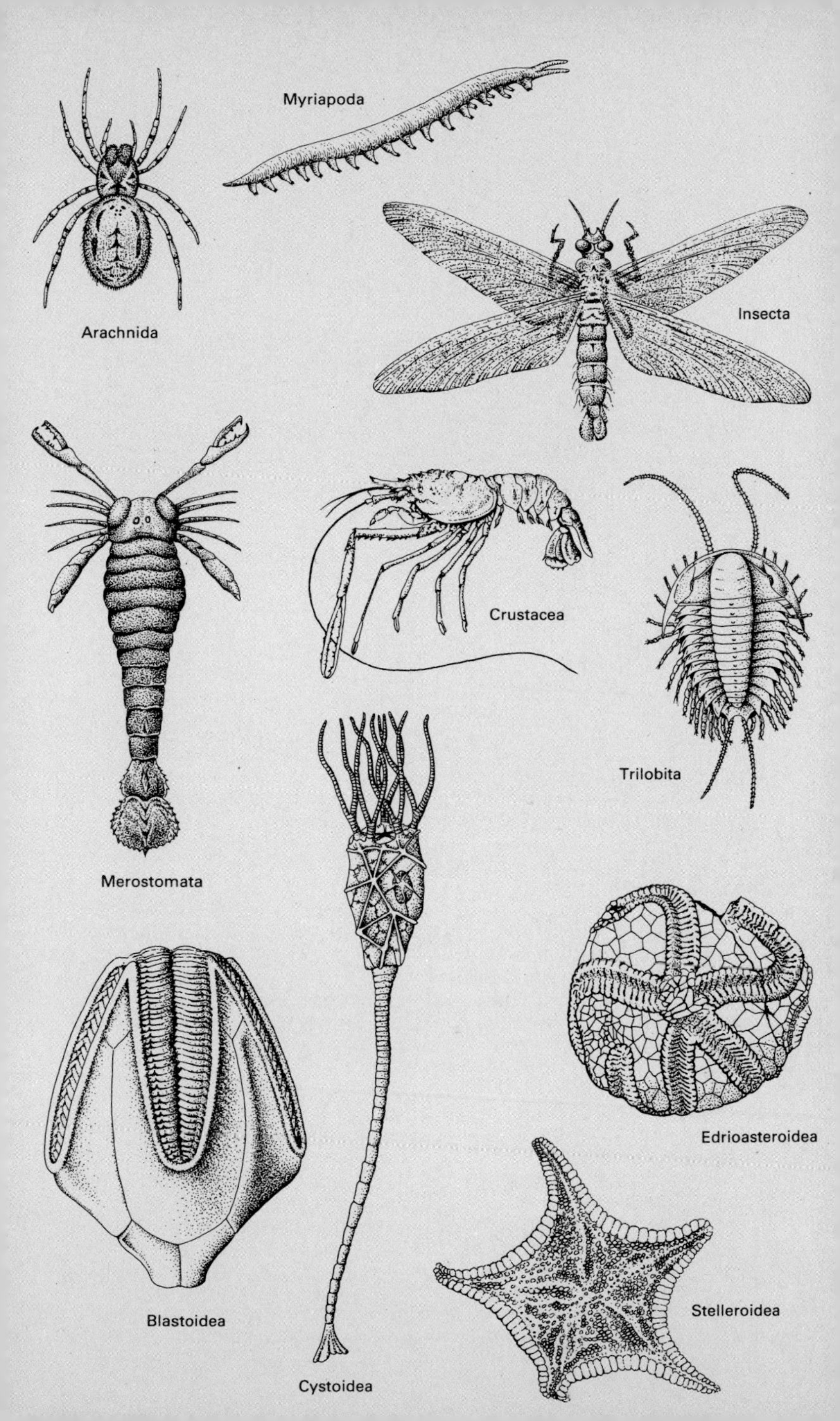
Myriapoda
Arachnida
Insecta
Merostomata
Crustacea
Trilobita
Merostomata
Edrioasteroidea
Blastoidea
Stelleroidea
Cystoidea

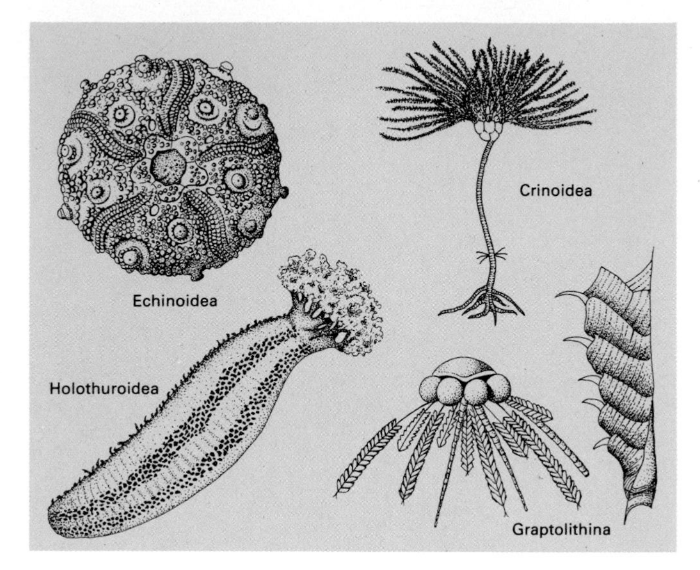

Echinoderms appeared during the Cambrian era and are still abundant in the seas of the modern world. They are subdivided into numerous classes, some of them highly important: Echinoidea, Crinoidea, Holothuroidea, Cystoidea, Edrioasteroidea, Blastoidea, Stelleroidea.

Phylum Hemichordata
This phylum has existed since the mid-Cambrian era and occurs today; these animals are characterized by the presence of a notochord and are regarded as being very close to the Chordata. They are subdivided into three classes:
Class Enteropneusta. These were first discovered in fossil form only a few years ago, although traces of their life on the sea floor were already known. They are animals of wormlike appearance, composed of a proboscis, a collar, a trunk and a tail.
Class Pterobranchia. Pterobranchs are colonial organisms arranged along an internal structure called a "stolon." Examples of the class have been discovered in Ordovician sediments.
Class Graptolithina. Now extinct, graptolites have been grouped under the phylum Hemichordata because of the affinities with pterobranchs displayed by these colonial Paleozoic animals; their chitinous colonies, in fact, display an analogous specialization.

Subphylum Vertebrata
Vertebrates belong to the phylum Chordata; they are equipped with a skeletal apparatus, either cartilaginous or osseous, composed of a variable number of vertebrae. They possess a central nervous sys-

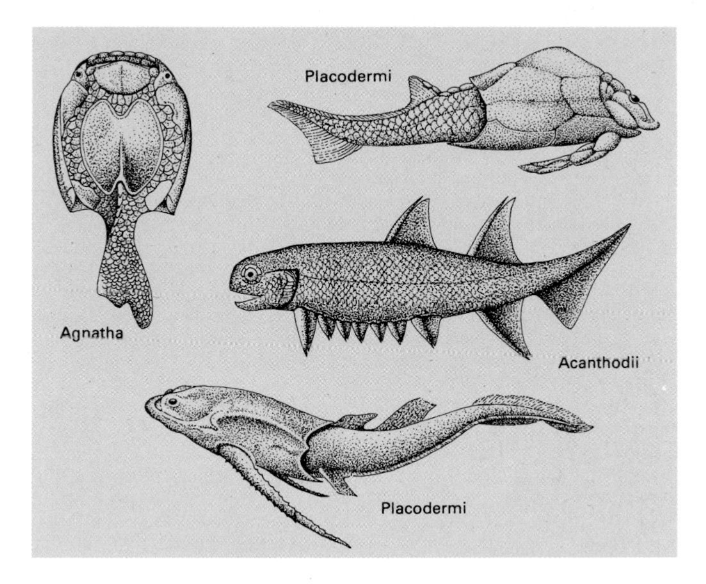

tem, which runs parallel to the spinal column and emerges into the cranium to form the encephalon. They are equipped with a respiratory system that has one part, the anterior portion, in common with the digestive apparatus, and their blood contains hemoglobin. Primitive vertebrates first divided into five classes, all of them fish (Agnatha, Placodermi, Acanthodii, Chondrichthyes, Osteichthyes), with amphibians, reptiles, birds and mammals arising later.

Class Agnatha. Agnatha are fish that first appeared in the Upper Cambrian era and survive today (they are, in fact, represented by the Cyclostomata); they are bereft of limbs and lack jaws. The oldest and most primitive vertebrates are members of this class, known as ostracoderms because of the stout armor in which they were covered. Ostracoderms, which became extinct in the Devonian, lived in both marine and fresh waters.

Class Placodermi. More evolved than Agnatha, the placoderms possessed jaws, paired forelimbs and hind limbs, and strong armor. The head was equipped with large eyes and the overall organization of the body was similar to that of sharks.

Class Acanthodii. These were the first fish to be endowed with jaws; they had a long, fusiform body and lived from the Silurian to the Lower Permian.

Class Chondrichthyes. Chondrichthyes are represented today by sharks and rays; they are fish with a cartilaginous spinal column and some paleontologists believe they are direct descendants of the placoderm arthrodires. Forms similar to modern sharks appeared in the Devonian.

Class Osteichthyes. Osteichthyes are fish that possess an osseous skeleton. Originating in the Silurian era, they are subdivided into three groups: Dipnoi, Actinopterygii, which includes the bony fish found in modern seas, and Crossopterygii, a group of fish that marked a major evolutionary stage; from one of its stock there emerged, during the Devonian, the tetrapod amphibians.

Class Amphibia. Amphibians are vertebrates that occupy an intermediary position between fish and reptiles. Within this class there are three distinct subclasses: Labyrinthodonta, Leptospondyli and Lissamphibia. The labyrinthodonts populated the lands that emerged between the Upper Devonian and Lower Jurassic. They were generally very large, up to 3 meters (9.75 ft) in length. The Leptospondyls appeared in the Lower Carboniferous period and became extinct at the end of the Permian; their dimensions were smaller. The Lissamphibia contain the still-extant amphibians, Urodeles, Anurans and Apoda, which appeared in the Middle Jurassic, the Lower Jurassic and the Paleocene eras, respectively.

Class Reptilia. The reptiles were widespread on earth during the Mesozoic era; but they appeared in the Carboniferous period, derived from a group of Amphibia. Unlike the amphibians, most detached themselves completely from the aquatic environment following the production of the amniotic egg which, possessing a shell, could also be laid on dry land. Reptiles first conquered the newly emerged land, then the air by evolving flying reptiles and then the marine environment by evolving swimming reptiles. The systematic subdivision of reptiles into six subclasses is based on the characteristic possession or absence of certain bones of the cranium to form the temporal fenestrae.

Subclass Anapsida. The anapsids consist of Cotylosaurs and Mesosaurs, now extinct, and Chelonia. They are characterized by the absence of temporal fenestrae and embrace the most primitive forms of reptile, in particular the Cotylosaurs who lived from the Carboniferous to the Triassic eras and have come to be regarded as the ancestors of the other reptiles.

Subclass Lepidosauria. Primitive diapsids (they possess two temporal fenestrae) whose modern representatives are rhynocephalians and Squamata (lizards and snakes). They first appeared during the Permian era and were very widespread in the Mesozoic.

Subclass Archosauria. Archosauria are characterized by a diapsid skull, but they are much more highly evolved than Lepidosauria, from whom they probably derived. Among the main differences between them is a tendency toward bipedality. Archosauria include the famous dinosaurs, flying reptiles and modern-day crocodiles.

Subclass Euryapsida. Euryapsids are a group of extinct reptiles abundant during the Mesozoic that possessed a single temporal fenestra in the upper section of the cranium. This subclass embraces a group of reptiles adapted to an amphibious or exclusively marine existence. Among its most outstanding marine representatives were the plesiosaurs, which became extinct in the Upper Cretaceous period, and the Placodontia. These creatures spent their lives mainly in water, even though they were still tied to terra firma for breeding and for resting when not hunting.

Subclass Ichthiopterygia. Ichthiopterygia appeared in the Triassic

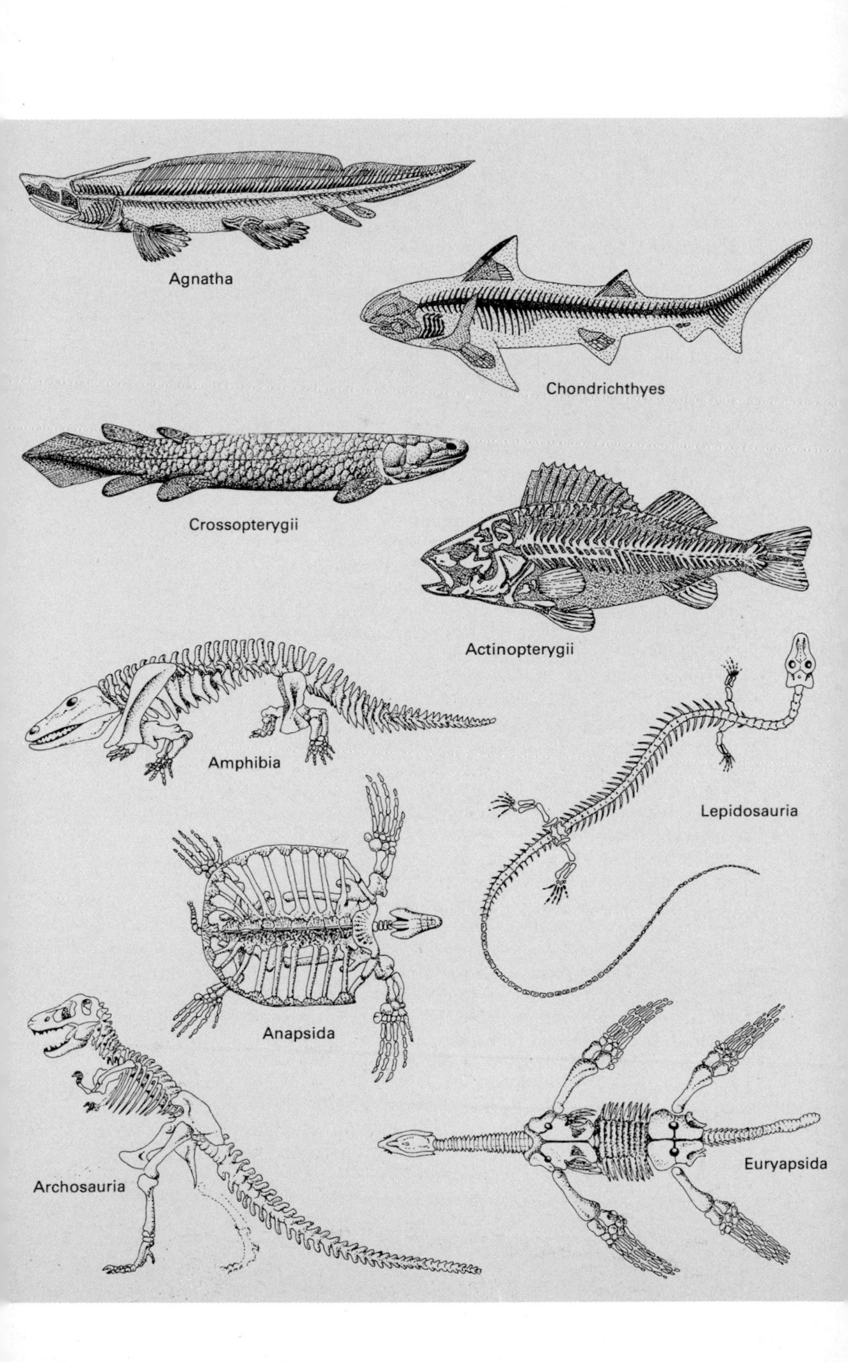

Agnatha
Chondrichthyes
Crossopterygii
Actinopterygii
Amphibia
Lepidosauria
Anapsida
Archosauria
Euryapsida

era in the form of the Ichthyosaur and became extinct by the Upper Cretaceous period. They were reptiles perfectly adapted to marine life, possessing a stream-lined body very similar to that of a dolphin and giving birth to living young in the sea. They possessed limbs adapted for swimming, particularly in the case of the front ones, on which the hand became a true paddle.
Subclass Synapsida. Synapsids are reptiles characterized by a temporal aperture placed in the lower part of the cranium. They are highly important in the evolutionary process, since it is believed that they gave rise to mammals through a series of modifications; it is for this reason that they are also known as mammal-like reptiles. They represent an extinct group that lived from the Permian to the Triassic eras.
Class Aves. The fossil record of birds begins in the Upper Jurassic. The first birds displayed reptilian characteristics mixed with typically avian features: in fact, they possessed a cranium and a long tail similar to that of the bipedal Saurischian dinosaurs from which they derived, as well as wings with feathers and pinions. The greatest period of avian development occurred in the Cenozoic era, when large numbers of different kinds of bird crowded the air.
Class Mammalia. Mammals are warm-blooded vertebrates whose body is covered in fur or hair and which are adapted to terrestrial, aquatic or aerial life. The earliest mammals have been discovered in rocks of the Triassic period; they were very small-sized animals, of similar dimensions to a mouse, and had no specialized diet.

Mammals are subdivided into three groups: monotremes (which reproduce by means of eggs), marsupials and placentals (which give birth to live young). In all mammals the newborn are fed on their mother's milk.

There are no paleontological data for monotremes prior to the Pleistocene, even though they display certain very archaic characteristics, such as the laying of eggs, a feature typical of reptiles (from which mammals are derived), whereas we know for sure that the divergence between marsupials and placentals took place during the Cretaceous era, with Pantotheres as the intermediary group.

Although they existed during the Mesozoic, mammals had to wait until the Tertiary for their great evolutionary explosion, when, following the disappearance of the dinosaurs, they replaced them in every environment. The most abundant group nowadays are the placentals, while the marsupials are typical of Australia, New Guinea and southern America, and monotremes are confined solely to Australia and New Guinea.

Below: Slab of Eocene sandstone bearing the imprint of a gastropod's trail.

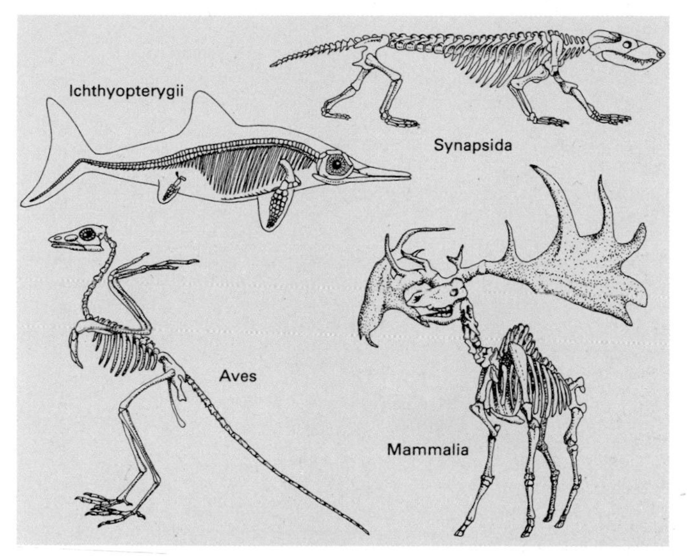

TRACE FOSSILS

Trace fossils are the imprints or markings left behind by living organisms. The term "trace fossil" is used generally to denote the footprints, trails, burrowings or any other type of mark resulting from the various activities of these organisms during life and preserved, whether in a soft sediment or hard substratum. The branch of science dealing with this phenomenon is called "ichnology," a word derived from the Greek words *ichnos*, meaning "imprint," and *logos*, meaning "study."

Trace fossils are found principally in sediments of marine origin, although there are abundant traces left by terrestrial animals, with the earliest examples dating back to the Precambrian. They are of considerable interest: even though they are clearly more vulnerable to destruction than the hard parts of an animal, due, among other things, to the action of tides and currents, their presence in many sedimentary deposits is of great help to scientists since they are preserved *in situ*, meaning that they have not been subjected to any movement and are therefore good indicators of the conditions prevailing on the seabed at the moment of their formation. In addition, biogenic sedimentary structures often escape the destructive processes to which the physical remains of the organisms' bodies are subjected; this means that in many rocks they are the sole evidence of past life.

The exact interpretation of trace fossils calls for an accurate comparison with the traces left by contemporary organisms. For example, it is very hard to identify an invertebrate that has a trace fossil, since different animals may leave very similar, if not identical, traces. Things are altogether different in the case of trails left by vertebrates, however; these often allow for the identification of the shape of the animal that made them and also for the latter's inclusion in a well-defined ecological group.

The systematic classification of trace fossils is a somewhat complex matter, since different trace fossils may be the result of different actions by the same animal and, in addition, may also depend on the characteristics of the stratum in which the traces are impressed. The best criterion for classification, therefore, would seem to be ethological, meaning one based on the type of action or behavior that has produced the imprint. Five main groups of structures can be recognized in this way: the dwelling structures of infaunal animals (*Domichnia*); structures due to the ingestion of mud by deposit-feeding organisms (*Fodinichnia*); traces of movement such as trails, burrows and tracks (*Repichnia*); traces left by byssally attached animals (*Cubichnia*). These groups leave traces that have characteristic shapes, regardless of the systematic position of the animal that produced them, since, as has already been mentioned, the same type of imprint can be produced by unrelated organisms.

As with organisms, trace fossils are subject to a binomial system of nomenclature, even though in this case the concepts of genus and species are endowed with a completely different significance.

A further two examples of trace fossils: Imprints of a seymouriamorph amphibian (above); rock with fucoids, worm tracks or plants (below).

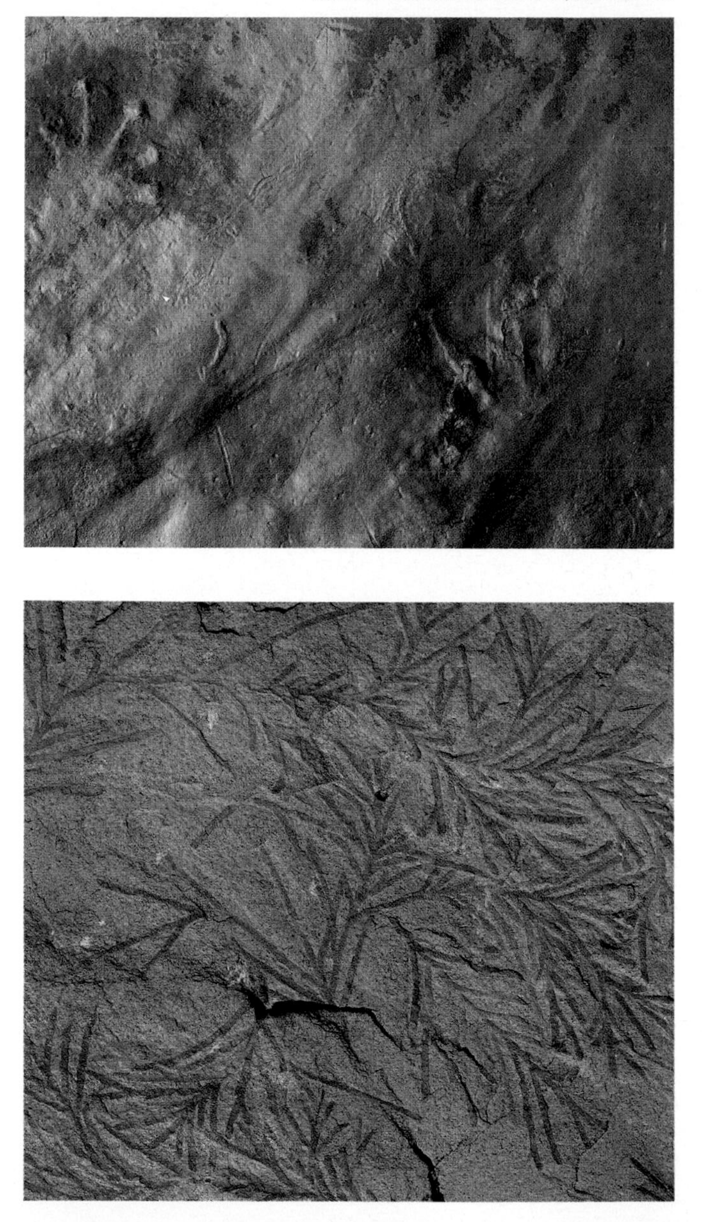

FOSSILS

1 JACUTOPHYTON

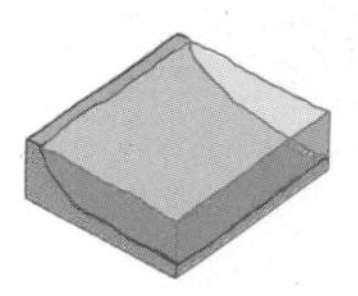

Classification Phylum Cyanophyta.
Description A genus of stromatolites (domed, pillarlike structures composed of superimposed layers). Stromatolites cannot really be regarded as the organic remains of particular organisms, but as calcareous or siliceous sedimentary structures produced by microorganisms. They are, in fact, the result of the growth of colonies of organisms, mainly blue-green algae. They vary in size from fractions of a micron to dozens of meters, with a similar wide variation in shape. The presence of laminae, however, is universal, and these are sometimes as much as several millimeters (less than a quarter of an inch) thick.
Stratigraphic position and geographical distribution The chronological distribution of stromatolites is vast. The earliest structures date from more than three billion years ago, and there are major calcareous deposits still in the process of formation, due mainly to the action of algae; among the most important examples are those in the marshy Everglades of Florida and the marine stromatolites in Shark Bay, Western Australia. The photograph shows the cross section of a stromatolite of the genus *Jacutophyton*; this particular specimen comes from Mauretania and dates from the Upper Precambrian (700 million years ago).
Note Stromatolites reached their acme during the late Precambrian, after which they went into a slow decline, possibly due to the appearance of more highly evolved algae and Metazoa, or when snails evolved, which feed on algae.

---

2 GYROPORELLA

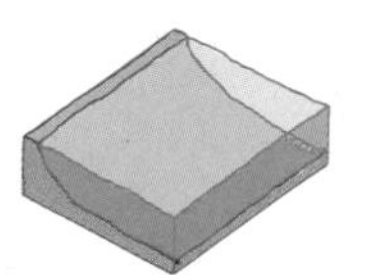

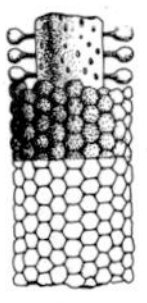

Classification Phylum Chlorophyta, Family Dasycladaceae.
Description A calcareous alga, characterized by the presence of a central thallus from which radiates a series of branches with swollen tips that combine to give the surface a granular appearance.
Stratigraphic position and geographical distribution The genus is known to have existed during the Middle and Upper Triassic (220–190 million years ago), and is particularly abundant in Triassic sediments in the Alps, from which the sample in the photograph is taken.
Note The family Dasycladaceae has been known since the Ordovician and there are still representatives living in temperate and warm seas at great depths, generally below 10 meters (32 ft). These algae are very abundant in fossil form owing to their ability to fix calcium carbonate, with which they constructed calcareous skeletons that have a good chance of surviving through time. Dasycladaceae were particularly abundant during the Triassic period, when they played a very important role in building great organogenic rocks like those found in the Italian Dolomites. The fossil forms of Dasycladaceae had a lifestyle similar to that of their modern counterparts.

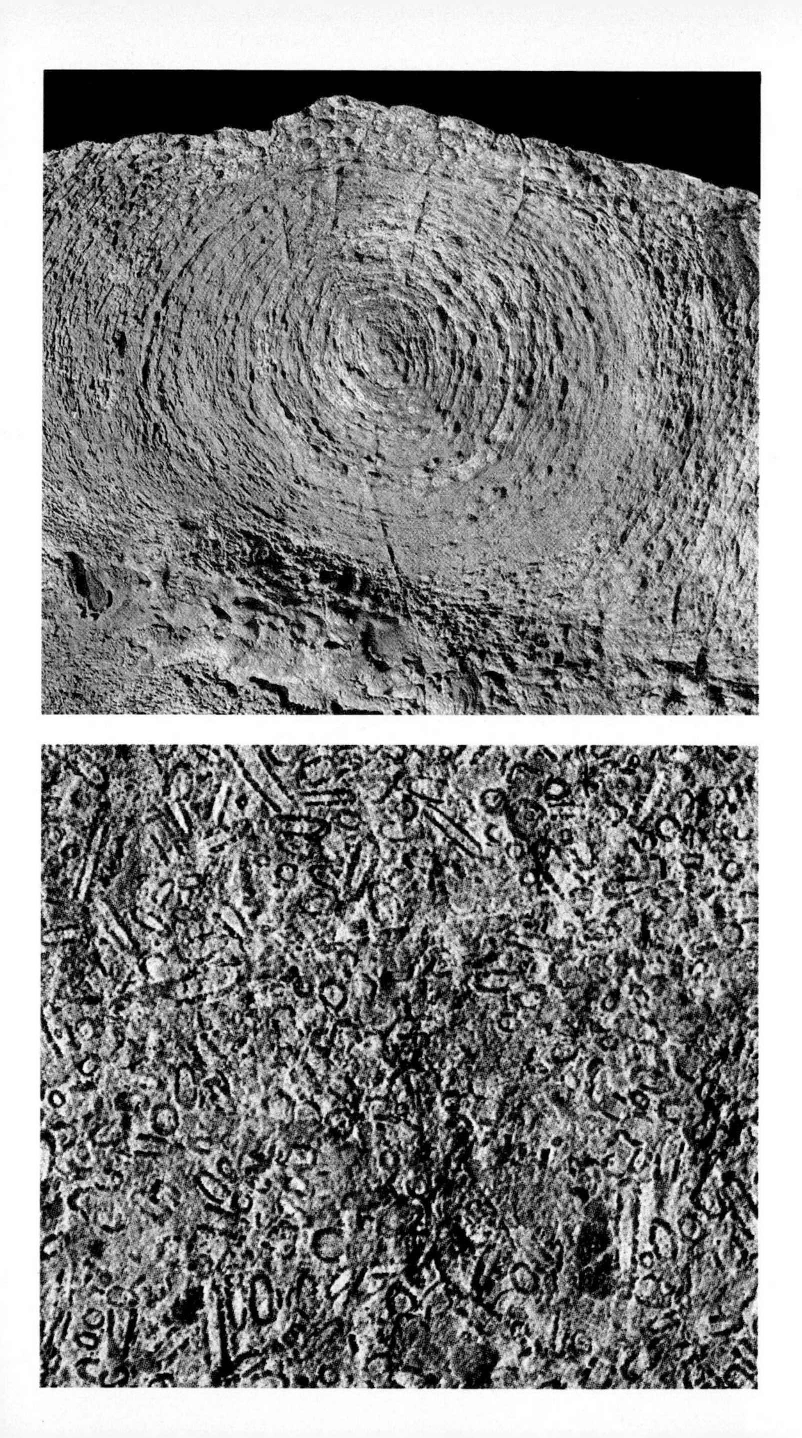

3 PSILOPHYTON

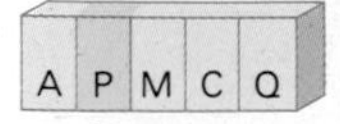

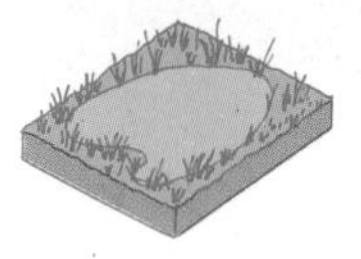

Classification Phylum Tracheophyta, Class Trimerophytopsida, Order Psilophytales, Family Psilophytaceae.
Description A very primitive vascular plant whose branches, bereft of leaves, subdivide mainly dichotomously, but also display lateral ramifications. The stems often possess spiny hairs scattered randomly over the surface. The fertile branches bear bunches of hanging, elongate ellipse-shaped sporangia (the receptacle containing the spores).
Stratigraphic position and geographical distribution The genus ranges from the Lower to Middle Devonian (395–360 million years ago) and is probably present in fossil form in rocks from the base of the Upper Devonian. Its geographical distribution is fairly broad, with different species of *Psilophyton* occurring in Canada, certain locations on the east coast of the U.S. and in Wyoming, in Europe (Germany, Bohemia, Norway) and in Asia. The example in the photograph, from the German Lower Devonian, measures approximately 12 centimeters (5 inches).
Note Plants similar to *Psilophyton*, the Trimerophytes, existed solely from the Lower to the Upper Devonian and display characteristics in common with primitive Rhyniopsida, from whom they are derived. The Trimerophytes appear to be the origin of the more evolved vascular plants that developed the apparatus of branches and leaves.

---

4 BARAGWANATHIA

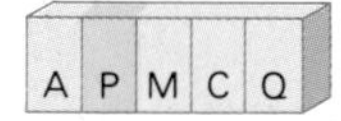

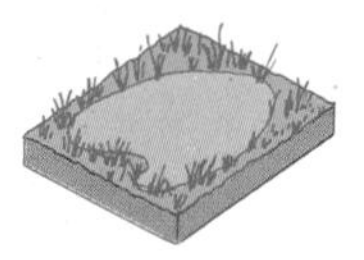

Classification Phylum Lycophyta, Class Lycopsida, Order Drepanophycales, Family Drepanophycaceae.
Description One of the two genera attributed to Drepanophycales, known essentially from fragments of branches with leaves measuring up to 28 cm (11 in) long. These branches are normally 1–2 cm (½–¾ in) wide, with dichotomous ramifications (the main axis is split into two secondary axes, which may themselves be further divided). The leaves, which entirely cover the youngest branches, are of simple, elongate form, up to 1 mm wide and 4 cm (1½ in) long. Their width remains virtually constant right to the tip, which may be either pointed or rounded.
Stratigraphic position and geographical distribution The species *B. longifolia* (see photograph) has been discovered in Upper Silurian strata (400 million years ago) in Victoria, Australia. Another species, *B. oehleyi*, based on specimens whose identification is still open to debate, dates from the Silurian and has been found in eastern Thuringia, Germany.
Note *Baragwanathia* is the oldest vascular land plant yet to be clearly identified as such, displaying conductor tissues and spores, and is also the oldest known lycophyte. Its Silurian age has been doubted because the guide fossil used to date the sediments in which it was found, the graptolite *Monograptus*, was in fact from the Devonian period. Further examples of *Baragwanathia* have nevertheless been discovered in sediments beneath Devonian strata and attributed to the Silurian on the basis of associated invertebrates.

5 PROTOLEPIDODENDRON

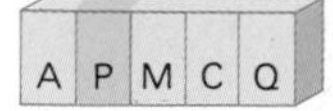

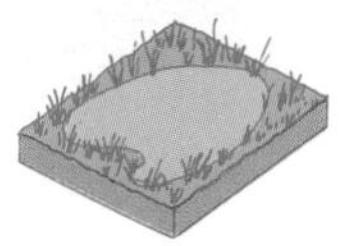

Classification Phylum Tracheophyta, Class Lycopsida, Order Protolepidodendrales, Family Protolepidodendraceae.
Description A plant with dichotomous stems that reach about 1.5 cm (½ in) in diameter. It possesses leaves with bifurcating tips, which are spirally arranged on the rhizomes and on the erect, dichotomous branches. The bases of the leaves, fusiform in shape, completely cover the surface of the branches, while the leaves themselves are arranged on the stems in a gently angled spiral. As a rule, the genus takes the form of a small herbaceous plant, of which only five species are known, all of them extinct.
Stratigraphic position and geographical distribution The genus *Protolepidodendron* is known from the Lower and Middle Devonian (370–360 million years ago) in Europe, U.S.S.R., China, Australia and North America. The example in the photograph, which belongs to the type species *Protolepidodendron scharyanum*, comes from the Middle Devonian (370–360 million years ago) in Australia; examples of the same species occur in Europe (Belgium, Germany, Czechoslovakia), U.S.S.R., China and the state of New York.
Note The leaf scars of the fallen leaves are shaped like small, elongated oval cushions and represent the first stages in the development of the leaf scar that later appears in well-developed form in the more evolved Lycopsida such as *Lepidodendron*.

---

6 LEPIDODENDRON

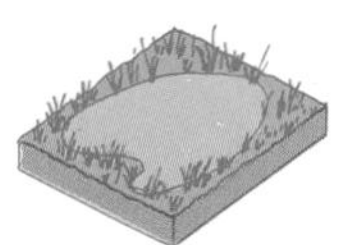

Classification Phylum Tracheophyta, Class Lycopsida, Order Lepidodendrales, Family Lepidodendraceae.
Description Arborescent lycopod that could normally reach a height of 30 meters (100 ft) with a base diameter of at least 30 cm (12 in); some examples possess a trunk some 35 meters (115 ft) long, up to the distal ramification, with a branching section continuing for a further 6 meters (20 ft) or so; reasonable estimates suggest that these trees could attain heights of more than 50 meters (16 ft) with a base diameter of more than 1 meter (39 in). The stigmarian roots stretched for at least 12 meters (39 ft) through the marshy substratum. Lepidodendrales achieved these dimensions within a few years, owing to the favorable climatic conditions in which they lived. No living species of *Lepidodendron* is known today, but the genus has been the subject of several studies. Some 110 species have been described on the basis of fragments of the outer cortex, which bears the leaf scars, and of mineralized specimens.
Stratigraphic position and geographical distribution *Lepidodendron* ranges from the Lower Carboniferous to the Lower Permian (345–270 million years ago), but some disputed remains date back to the Lower Devonian (390 million years ago). It occurs in numerous strata in Europe, North Africa, U.S.S.R., China and Mongolia; one species from the late Paleozoic has also been found in Patagonia.
Note *Lepidodendron* was a main inhabitant of the damp forests of the Upper Carboniferous, to judge from its abundance in strata of that period.

7 SIGILLARIA

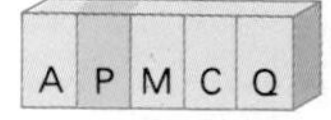

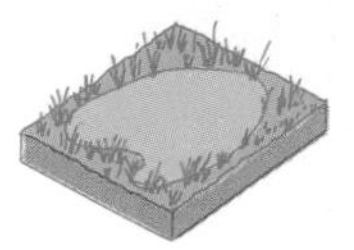

Classification Phylum Tracheophyta, Class Lycopsida, Order Lepidodendrales.
Description The absence of repeated ramifications at the apex of the trunk and the structure of the leaf bases left on the bark by the fallen leaves are the main characteristics distinguishing the genus *Sigillaria* from other Lepidodendrales. The genus includes arborescent forms which, although some could reach heights of 34 meters (111 ft), were generally less than 20 meters (65 ft) tall. The leaf bases of *Sigillaria* were arranged helicoidally on the trunk and in fossil form show a precise vertical alignment. The leaves are lanceolate or linear, longer and broader than those of *Lepidodendron*, up to 1 meter (39 in) long and 1 cm (½ in) across. A *Sigillaria* plant would have looked highly unusual to us, its very short branches and leaves closely bunched together in large clumps at the distal extremity of the trunk. Isolated leaves and fructifications, attributed to a variety of genera, and isolated stigmarian root section, are all that have been discovered of *Sigillaria*.
Stratigraphic position and geographical distribution The genus ranges from Lower Carboniferous to Permian (345–235 million years ago) most abundant in Europe and N. America; the example shown is from the German Carboniferous.
Note With *Lepidodendron, Sigillaria* was one of the key arborescent forms in the marshy forests of the Carboniferous; during this period Lepidodendrales were so abundant that many layers of fossil carbon are composed solely of their remains.

---

8 STIGMARIA

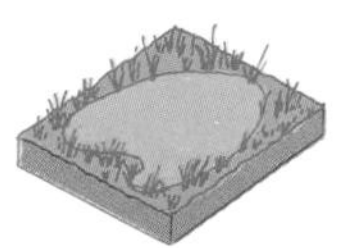

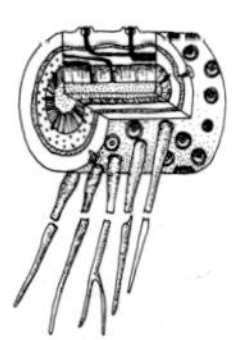

Classification Phylum Tracheophyta, Class Lycopsida, Order Lepidodendrales.
Description A genus established to describe the roots (rhizomes and rhizoids) of Lepidodendrales fossilized separately from the rest of the plant. They take the form of a stout, cylindrically shaped rhizome to which there are sometimes roots still attached, with the latter emerging at right angles from the rhizome; more often, however, only the rhizome survives, bearing the scars that mark the points at which the roots were inserted in a helicoidal arrangement along the rhizome.
Stratigraphic position and geographical distribution Roots attributed to *Stigmaria* range from the Carboniferous to the Permian (345–235 million years ago). The genus has a very wide distribution, corresponding to the distribution of such widespread genera of Lepidodendrales as *Lepidodendron* and *Sigillaria*. The example in the photograph belongs to the type species *S. ficoides* and comes from the German Carboniferous; the roots emerging from the rhizome can be seen clearly.
Note The rhizomes, or underground stems, of Lepidodendrales form large systems that in many of the giant arboreal forms developed almost horizontally and extended for many meters around the plant; there are notable examples of these root systems having survived *in situ* together with the base of the trunk, some of them from the English Carboniferous.

9 CALAMITES

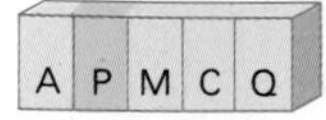

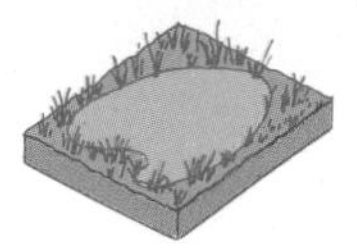

Classification Phylum Tracheophyta, Class Sphenopsida, Order Equisetales, Family Calamitaceae.
Description Arborescent Equisetum, 10 meters (32½ ft) long, sometimes reaching 20 meters (65 ft). The aerial stems sprout from nodes on the stout underground rhizome, which had adventitious roots at each node. The branches and leaves were joined on the subaerial stems in verticils matching the nodes. There are five subgenera based on the characteristics in the fossilized remains of the medulla: *Mesocalamites* is distinguished by the superficial parallel ridges of the vascular system alternating only partially after each node; in all other forms the ridges are invariably alternate; *Stylocalamites* has a few, irregularly arranged branches; *Crucicalamites* has branches at each node, making it bushy; *Diplocalamites* has a pair of branches at the nodes, opposite each other; and *Calamitina* has a regular ramification with many branches, but not at every node.
Stratigraphic position and geographical distribution. Widespread from the Middle Carboniferous to the Upper Permian (300–250 million years ago), *Calamites* occurs in many European strata and in the U.S.S.R., Asia Minor, China, Korea, Sumatra and the U.S. The example shown is from Illinois.
Note Very abundant in the warm marshy forests of the Upper Paleozoic. The genus was described on the basis of molds of medullae formed when the central cavity of the stem became filled with sediment: the mold was preserved in fossil state while the other tissues gradually decomposed.

---

10 ANNULARIA

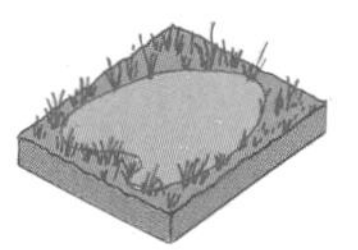

Classification Phylum Tracheophyta, Class Sphenopsida, Order Equisetales, Family Calamitaceae.
Description Genus that comprises the isolated fossil remains of the leaves of Calamitaceae. The leaves, either oval or lanceolate, are typically grouped together in verticils in varying numbers. Each leaf possesses a single, centrally placed nervature, which can occupy as much as half the width of the leaf itself. The length of the leaves varies, (in some forms reaching as much as 8 cm (3½ in), and is sometimes greater than the distance between the internodes, which, with their regular spacing, are characteristic of the stem of this plant genus; the ratio between the length and breadth of the leaves is approximately 8 to 1.
Stratigraphic position and geographical distribution The genus is very widespread in the Carboniferous (345–280 million years ago) in the U.S., Canada, China and Europe; it also occurs in the Permian in the U.S.S.R. and China and in the Upper Paleozoic in Patagonia. The example in the photograph comes from the Italian Carboniferous.
Note One typical feature of *Annularia* fossils is the arrangement of the individual leaf whorls, which usually lie on the same plane as the stem from which they sprout. This may be due to the deformation undergone during the fossilization process, but some authorities have suggested that this is a mirror-image of the original appearance of the plant, which was formed this way in order to expose the greatest possible surface area to the sunlight.

11 SENFTENBERGIA

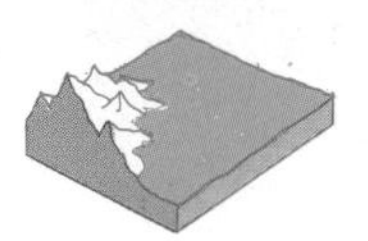

Classification Phylum Tracheophyta, Class Phyllicopsida, Family Schizaeaceae.
Description Leaves composed of pinnules with lateral edges that are either parallel or gently convergent, attached to the rachis for the full length of their base and very similar to those of *Pecopteris*, with the sporangia arranged in two lines along the lower edge of the pinnules. It possesses fronds larger than those of modern-day members of the same family; there are, in fact, incomplete fossil specimens some 40 cm (16 in) long and up to 30 cm (12 in) wide, while the sporangia are up to 1 meter (39 in) long.
Stratigraphic position and geographical distribution The genus *Senftenbergia* is known in different species in the European Carboniferous, the most notable being *S. plumosa*, to which the example in the photograph, from the German Carboniferous (300 million years ago), belongs; a variety of this species is also known from the Lower Carboniferous (Mississippian stratum) in Illinois. The earliest known species comes from the Lower Carboniferous in Musselburgh, Scotland.
Note *Senftenbergia* represents the earliest known form of the Schizaeaceae fern family, now found mainly in tropical or subtropical environments and comprising four modern-day genera, some of which, such as *Lygodium*, are also frequently found in fossil form in rocks of the Cretaceous and Tertiary.

---

12 SPHENOPTERIS

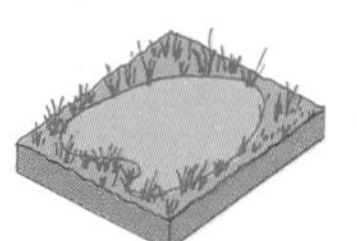

Classification Phylum Tracheophyta, Class Gymnospermopsida, Order Pteridospermales.
Description Genus of pteridosperm included in the Sphenopterid family, a complex and heterogeneous group, poorly defined, comprising herbaceous, shrub and small arboreal forms sharing certain common characteristics, particularly the morphology of their pinnules and their nervature. The general morphology, as is typical of pteridosperms, is that of a fern; the fronds consist of a principal axis subdivided into secondary ramifications with slightly elongate pinnules, whose incised outline forms rounded lobes. Insertion at the base is direct; the nervature displays a fan-shaped arrangement. Fertile fronds bearing the reproductive structures are also known.
Stratigraphic position and geographical distribution Typical form of Paleozoic pteridosperm, the genus *Sphenopteris* occurs in the Carboniferous, from the Namurian to the Westphalian levels (325–290 million years ago). It is very typical of the central-western European Carboniferous. The example in the photograph belongs to the species *S. elegans* and comes from the German Carboniferous.
Note The morphology of the entire plant is not known; small in size, it must have been typical of such damp Carboniferous environments as forests in marshes by the sea or fresh inland waters.

13 GLOSSOPTERIS

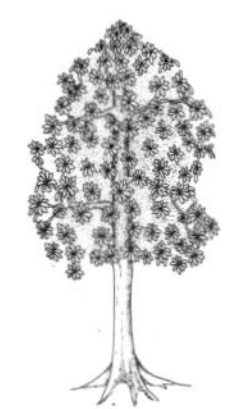

Classification Phylum Tracheophyta, Class Gymnospermopsida, Order Glossopteridales.
Description A typical pteridosperm, long believed to have been in herbaceous or bush form. It is, in fact, an arborescent form, which reached some 6 meters (20 ft) in height, with a stem 40 cm (16 in) in diameter, whose wood shows remarkable structural similarities with that of typical gymnosperms. The fossil trunks show clear growth rings. The leaves of *Glossopteris* represent the most common part preserved in fossil form. They are characteristically elongate and lanceolate, with the nervatures forming a very typical network of veins and with stamens borne on the underside of the leaves. The reproductive structures and fossil pollens of *Glossopteris* are also known.
Stratigraphic position and geographical distribution The genus is typical of the Permian and Triassic (270–190 million years ago) in India, Australia, South Africa, South America and the Antarctic, but some examples have been discovered in Jurassic sediments in Mexico. The example in the photograph is from the Antarctic.
Note At the end of the Paleozoic *Glossopteris* was so characteristic of the single southern supercontinent known as Gondwanaland (the continents of the Northern Hemisphere were united in a land mass known as Laurasia) that the Gondwana flora is also known as "*Glossopteris* flora." The surface area of Gondwanaland coincides exactly with the geographical distribution of *Glossopteris*.

---

14 NEUROPTERIS

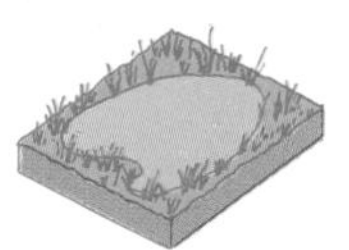

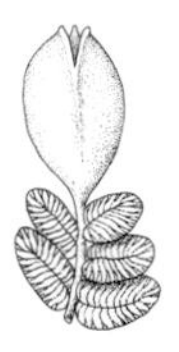

Classification Phylum Tracheophyta, Class Pteridospermopsida, Order Pteridospermales.
Description This genus was established to categorize a type of pteridosperm leaf found in isolated fossil form, detached from the original plant. These leaves can be large, with a rachis marked by longitudinal striations. In the unramified parts the main axis bears rounded pinnules; on the secondary ramifications there are large leaves, more elongate, with subparallel sides and narrower extremities. Leaves of the *Neuropteris* type are found on plants such as *Medullosa*, an arboreal form of pteridosperm, some 3.5 meters (12 ft) tall, whose leaves, when found in isolation, are attributed to a variety of genera, among them *Neuropteris.*
Stratigraphic position and geographical distribution *Neuropteris* is typical of the Upper Carboniferous (280 million years ago). It is very widespread in European and North American strata; the example in the photograph belongs to the species *N. gigantea* and comes from the Pennsylvanian stratum at Mazon Creek, Illinois.
Note Pteridosperms, known solely in fossil form, were very abundant in the Paleozoic and Mesozoic. They generally resembled tree ferns, although in some cases they recall climbing plants such as vines. Because of their similarity to ferns they were for a long time assigned to the vine group: only at the beginning of this century was it discovered that they represent a separate group whose resemblance to ferns is due simply to parallel evolution.

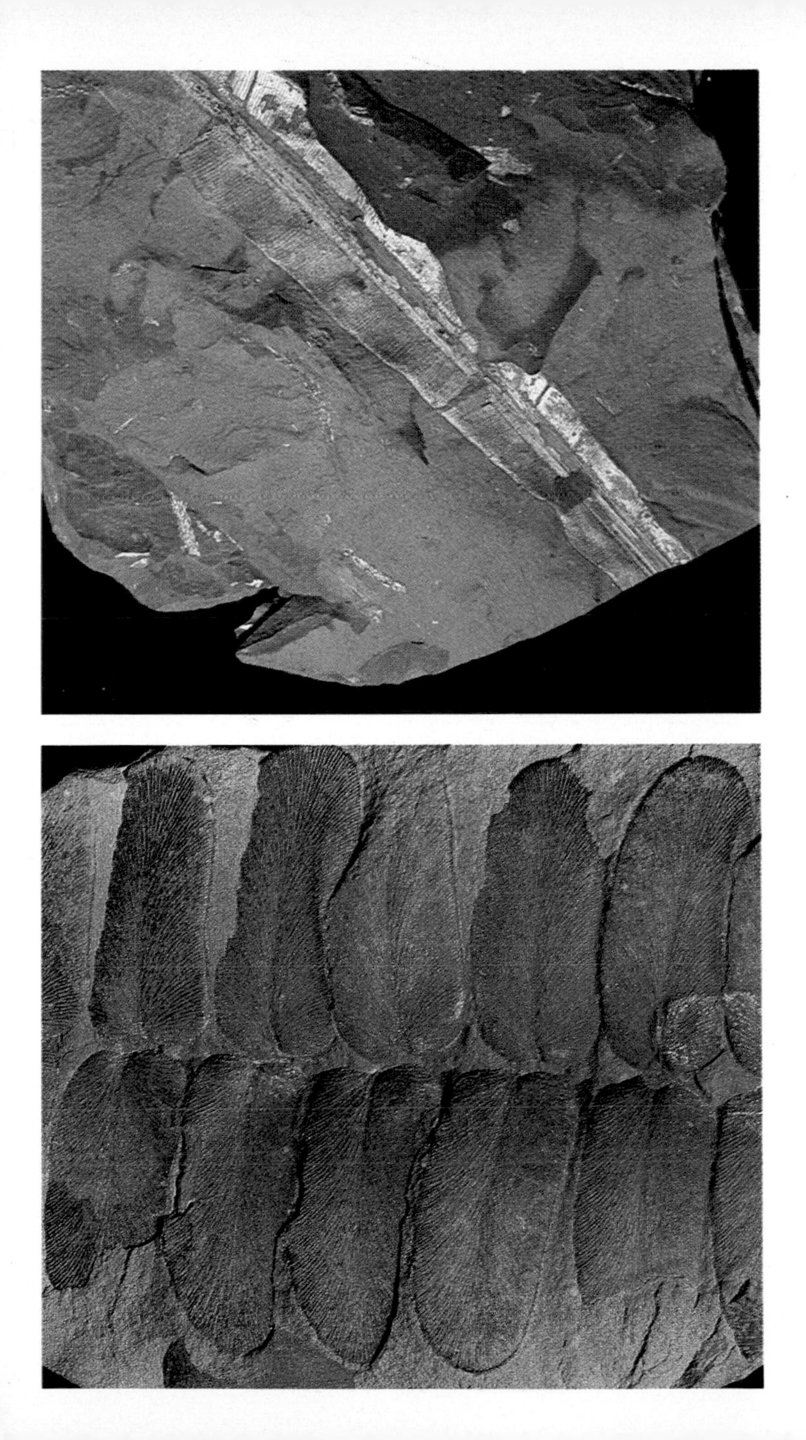

15 ZAMITES

Classification Phylum Tracheophyta, Class Gymnospermopsida, Order Cycadales.
Description A genus used to indicate the fossil remains of cycad fronds. These very distinctive leaves are of pinnate type, composed of a central axis from which emerge two rows of elongate leaves, lanceolate in shape and equipped with dichotomously dividing nervatures. The leaves assigned to the genus *Zamites* belong partly to the family Cycadeae, but leaves attributed to the same genus have been found on plants of the genus *Cycadeoidea*, which belongs to the family Cycadeoideae. The genus *Cycadeoidea* possessed a squat trunk, roughly cylindrical or columnar, about 1 meter (39 in) tall, at the tip of which there was a crown of pinnate leaves of the *Zamites* type.
Stratigraphic position and geographical distribution Leaves classified within *Zamites* range from the Triassic to the Lower Cretaceous (235–120 million years ago) and are very widespread in many areas of the world. The example in the photograph comes from the Lower Jurassic (Sinemurian, 190 million years ago) at Osteno, Lombardy (Italy).
Note Leaves constitute the most common fossil remains of cycad plants, which were very abundant in the Mesozoic era, when they reached their acme. They still survive in tropical forms, some of them arborescent, with characteristics similar to ferns or palms. Cycadeoideae, by contrast, which had leaves morphologically akin to those of Cycadeae, existed solely during the Mesozoic.

---

16 GINKGO

Classification Phylum Tracheophyta, Class Gymnospermopsida, Order Ginkgoales.
Description An arborescent form whose leaves have a distinctive fan or kidney shape, with a thin elongate stem, and are grouped together at the tips of the short branches or scattered along elongated branches. The shape and size of the leaves diverge considerably from the most typical bilobate form; the fast-growing leaves on the long branches display a deeply cut edge and are subdivided into wedge shapes, whereas the leaves on the short branches may have an unbroken outline. There is thus a certain variability within the leaves on a single plant, a fact which must be taken into account when classifying isolated fossil leaves. The distinctive forms of the leaves, however, combined with other peculiarities in their structure (the thin, dichotomously branching veins, for example), make it easy to distinguish the leaves of the genus *Ginkgo* from others.
Stratigraphic position and geographical distribution The genus *Ginkgo* is known since the Middle Jurassic (160 million years ago); it reached its greatest period of expansion during the Cretaceous and the early Tertiary periods, surviving today in a single species. The example in the photograph, which belongs to the species *G. huttoni*, comes from the Yorkshire Middle Jurassic.
Note Ginkgoales date back to the Permian period and today survive in a single species, *G. biloba*, discovered in 1956 in the wild state in China.

17 FUSULINA

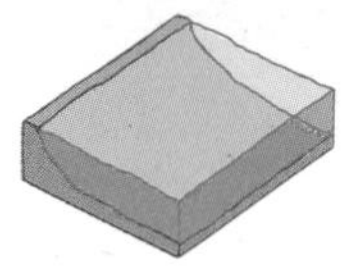

Classification Phylum Protozoa, Class Foraminifera, Suborder Fusulinina, Family Fusulinidae.
Description Fusulinae are exclusively marine invertebrates that possess a fusiform test wrapped along an axis that coincides with the maximum diameter of the shell. The wall of this test is formed of two layers, a thinner, external one (tectum) and a thicker, internal one (diaphanotheca); the combination of the two layers is called the wall or spirotheca. This bends rhythmically toward the whorl at the tip to form a series of chambers, themselves separated by septa. The dimensions of the test vary from 2 to 10 mm (less than .5 in) in length.
Stratigraphic position and geographical distribution The genus *Fusulina* is a guide fossil for sediments referable to the Upper Carboniferous (290 million years ago); the concentration of a large number of specimens whose calcareous tests form the organogenic part of certain limestones has led to the latter being known as "fusulinid limestone." The genus is distributed throughout the world. The examples in the photograph come from the Permian at Alma, Kansas (370 million years ago).
Note The family Fusulinidae is extremely important from a stratigraphical standpoint, since certain genera attributable to this family are used to establish the date of sediments deposited during the Carboniferous and Permian. From a paleoecological point of view the fusulinids indicate warm marine conditions in a continental shelf environment.

---

18 NUMMULITES

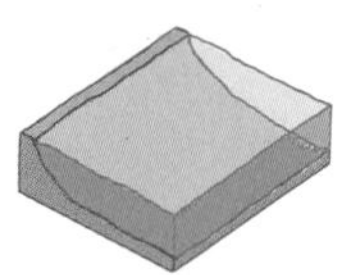

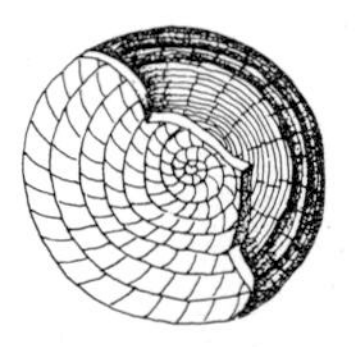

Classification Phylum Protozoa, Class Foraminifera.
Description Nummulites are disk-shaped Foraminifera with bilateral symmetry; the test is formed by a wall or plate which coils in a spiral around a central axis. This wall is calcareous, perforated and very robust, with, on the inside, a thin layer, similarly calcareous, which folds downwards to create septa forming individual chambers. There may also be a series of internal supports whose role is to further strengthen the test. The external appearance of the shell may be either smooth or finely granulated. The dimensions of these Foraminifera varies from a few millimeters to 7–8 cm (2¾–3¼ in).
Stratigraphic position and geographical distribution The stratigraphic position of *Nummulites* ranges from the Paleocene to the Middle Oligocene; they reached their greatest size during the Eocene. They sometimes occur in such large concentrations that they form entire rock units such as the "nummulitic limestone" outcrops found in Italy and other countries outside Europe. The photograph shows *N. gizehensis* from the area around Cairo in Eygpt, dating from the Eocene (50 million years ago).
Note All members of the family Nummulitidae are excellent guide fossils for the Tertiary, since they evolved very rapidly and are very widespread. The distinction of the genus *Nummulites* at species level is based on the external ornament, size and shape of the test and the arrangement of the septa. Internal characteristics can be observed only by removing the outer layer or making thin longitudinal or transversal sections.

19 SIPHONIA

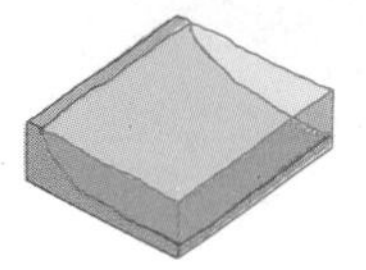

Classification Phylum Porifera, Class Demospongea, Order Lithistida, Family Halurhoidae.
Description A lobeless, pear-shaped sponge with a long pedicle whose ends, similar to tree roots, attach it to the sea floor. It possesses a pseudogastric central cavity and a rather complex and developed vascular system. The latter characteristic is typical of the family Halurhoidae, to which *Siphonia* belongs.
Stratigraphic position and geographical distribution The genus *Siphonia* ranges from the Middle Cretaceous to the Tertiary. Its area of distribution is restricted solely to Europe. The example in the photograph is a member of the species *S. piriformis* and comes from the English Upper Cretaceous (80 million years ago); it measures approximately 5 cm (2 in).
Note The class Demospongea is characterized by the presence of spicules, either siliceous or composed of spongin; sometimes both types may appear in the same sponge. The class dates from the Cambrian and still exists today.

---

20 RAPHIDONEMA

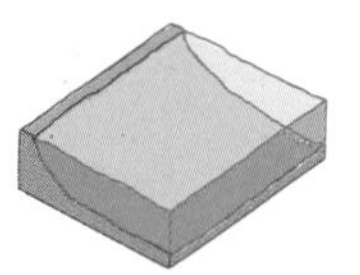

Classification Phylum Porifera, Class Calcarea, Order Pharetronida, Family Lelapiidae.
Description The general form of this sponge recalls that of a vase or broad funnel. Its external surface is characterized by the presence of lump-shaped swellings; the surface is generally very rough.
Stratigraphic position and geographical distribution The genus *Raphidonema* ranges from the Triassic to the Cretaceous and is confined to Europe. The example in the photograph, 13 cm (5¼ in) in diameter, comes from the Aptian (Lower Cretaceous, 115 million years ago) at Faringdon, Berkshire (England), and belongs to the species *R. farringdonense*.
Note The class Calcarea is characterized by sponges possessing skeletons composed of calcareous spicules. The class is known from the Cambrian onwards and still lives in marine waters today.

21 OCTOMEDUSA

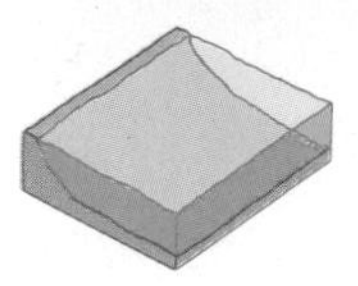

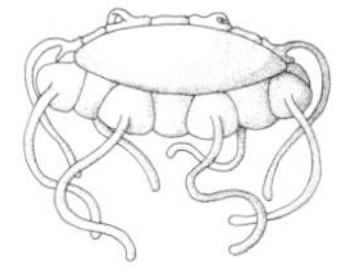

Classification Phylum Coelenterata, Class Scyphozoa, Order Coronata.
Description A jellyfish with a maximum diameter of a little more than 20 mm (¾ in) and a minimum of 3 mm (less than ¼ in), comprising a very small umbrella, round in shape and edged by a circular groove. Behind the groove is a lobate area from whence emerge eight tentacles whose length is equal to, or slightly greater than, the diameter of the umbrella. At the center of certain fossil examples one can see the mouth, quadripartite in form, and the manubrium, which is short. The different parts of the fossilized body are generally recognizable as a light-colored stain in the surrounding rock and, because of the insubstantiality of the body of a jellyfish, only rarely does the fossil stand out in relief.
Stratigraphic position and geographical distribution The genus is known from a single species, *O. pieckorum*, in the Pennsylvanian stratum (Upper Carboniferous, 300 million years ago) at Mazon Creek, Illinois, an example of which can be seen in the photograph; it measures approximately 1.5 cm (½ to ¾ in).
Note Examples of *Octomedusa*, like the other Mazon Creek fossils, are preserved in elliptical or rounded nodules of siderite, the formation of which appears to have been fostered by the presence of the organisms' remains in the sediment. Jellyfish of this genus are often found in groups in these nodules, as though they had been carried along by the currents, like jellyfish today.

---

22 RETICULOMEDUSA

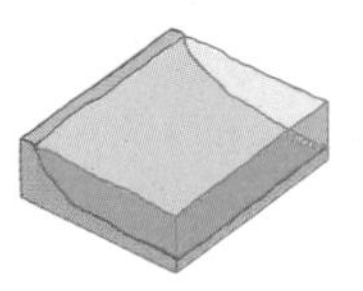

Classification Phylum Coelenterata, Class Schyphozoa, Family Rhizostomeae.
Description A small jellyfish with a maximum diameter of approximately 45 mm (1¾ in), with a circular umbrella around which there develops a lobate fringe; fossils show the development of a lobe of unusual dimensions on only one side. In fossils, the circular area of the umbrella takes the form of a fairly prominent dome and displays a distinctive reticulation due probably to the conformation of the gastrovascular apparatus, whose development recalls that of modern-day Rhizostomeae; there is also a quadripartite mark that corresponds to the mouth.
Stratigraphic position and geographical distribution Like *Octomedusa*, the genus *Reticulomedusa* is known in the Pennsylvanian stratum (Upper Carboniferous, 300 million years ago) of Mazon Creek, Illinois, in the species *R. greenei*, to which the example in the photograph, measuring 2 cm (¾ in), belongs.
Note The more highly developed lobe displayed by this fossil at the edges of the umbrella is not easy to interpret: it probably marks the spot where the gas produced by the decay of the organism's body built up, or even the place where the internal structures accumulated at the moment of the organism's death.

23 ESSEXELLA

Classification Phylum Coelenterata, Class Scyphozoa, Family Rhizostomeae.
Description A genus of medium-sized Scyphomedusae, varying from 1 cm (½ in) to 19 cm (7½ in) in length and from 6 mm (¼ in) to 9 cm (3½ in) in width. Its shape is generally elongate and, seen from the side, may be elliptical or almost either square, rectangular or triangular, although mainly elliptical or subrectangular with one of its extremities recalling the cap of a mushroom and corresponding to the umbrella of the jellyfish. This extremity can occupy anything from a sixth to a half of the animal's overall length; on the whole, however, its length is equivalent to roughly a third of the total body length. Only in rare cases will this section be either barely discernible or completely absent. The remaining portion of the animal displays a structure shaped like a skirt, which was probably a membrane that hung from the creature's umbrella.

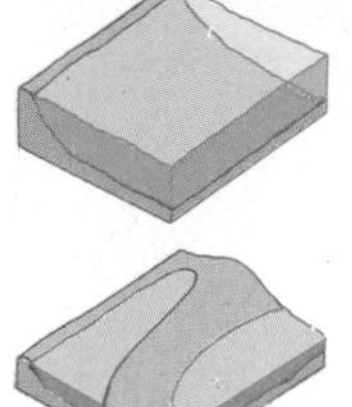

Stratigraphic position and geographical distribution The type species, *E. asherae*, shown in the photograph, is known in the Pennsylvanian stratum (Upper Carboniferous, 300 million years ago) of Mazon Creek, Illinois.
Note The skirt-like membrane encloses other structures that hang from the umbrella, and when the jellyfish fossilizes, the membrane covers the inner structures, so that proper observation of them is impossible, making a reconstruction difficult. It is, however, possible to make out up to eight lobes or tentacles.

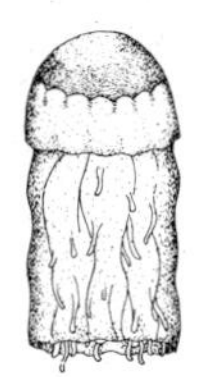

24 RUGOCONITES

Classification Phylum Coelenterata, Class Scyphozoa, Order Brachinida, Family Brachinidae.
Description These take the form of circular or slightly oval impressions, 6 cm (2½ in) or more in diameter. The imprint shows a number of radial canals diverging from a small, spherical dome at the center, which is slightly raised; these canals may be dichotomously subdivided, either once, twice or three times.

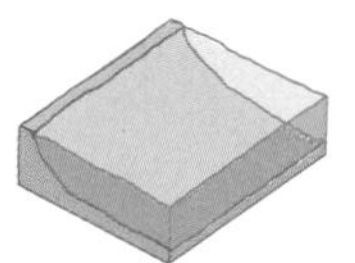

Stratigraphic position and geographical distribution The genus *Rugoconites* is known from the Upper Precambrian of the famous Ediacara beds in South Australia (670 million years ago), from which come two distinct species. The example in the photograph, which measures approximately 6 cm (2½ in), belongs to one of these, *R. enigmaticus*.
Note The genus *Rugoconites* is one of the oldest known metazoans. The Ediacara beds, from which it comes, is the place where fossils of multi-celled organisms in Precambrian rocks were first discovered. Before the Ediacara finds, which occurred during the years immediately after World War II, no organisms of this type had been definitely identified in rocks earlier than the Cambrian, which marks the beginning of the Paleozoic era.

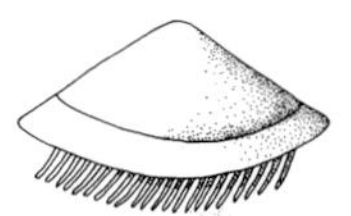

25 CHARNIODISCUS

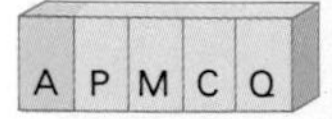

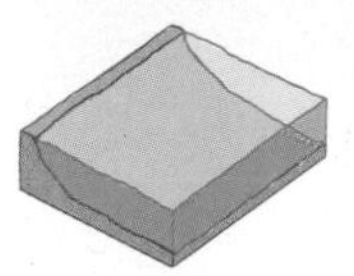

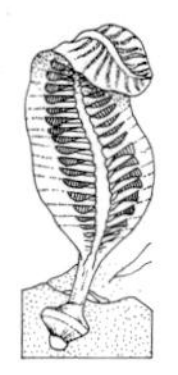

Classification Phylum Coelenterata, Class Anthozoa, Order Pennatulacea, Family Charniidae.
Description This is one of Ediacara's most fascinating Precambrian fossils: it consists of a central axis from which there emerge leaflike structures; the axis is anchored to a supporting disk. On the leaves are structures in which the polyps lived, and along the whole length of the fossil there are small rectilinear depressions that have been interpreted as spicules.
Stratigraphic position and geographical distribution The genus occurs in the famous Ediacara beds in South Australia, dating from the Upper Precambrian (670 million years ago) and several species have been recognized within the same horizon. The example in the photograph belongs to the species *C. oppositus* and measures 11 cm (4½ in).
Note Since the discoveries made at Ediacara, this type of fossil has been found in other Precambrian strata in the Northern Hemisphere: it displays significant similarities with the living coelenterate Pennatulacea, which, as the name suggests, resemble bird feathers in appearance. They are colonial organisms in which the individual component has a subcylindrical shape and generates around it other organisms that ultimately give the colony its characteristic appearance. The colony lives affixed to the seabed, in mud or sand, at a variety of depths; the Precambrian Pennatulacea must have had a similar lifestyle.

---

26 CALCEOLA

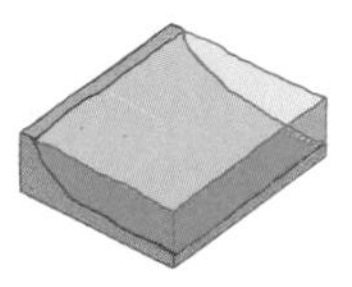

Classification Phylum Coelenterata, Class Anthozoa, Order Rugosa, Family Goniophyllidae.
Description The genus *Calceola* is represented by solitary corals with a skeleton whose shape recalls that of a sandal (a typical example is *C. sandalina*). The corallite is semicircular and the calyx is closed by an operculum; the calyx is inserted in the dorsal side, which is flattened and carries the hinge of the operculum. Upon removing the operculum one notices the presence of a cardinal septum which matches a small fossula accompanied by a large number of secondary septa.
Stratigraphic position and geographical distribution The genus *Calceola* is known in strata of the Lower and Middle Devonian in Europe, especially Germany, and Asia. It has also been discovered in the Middle Devonian of Africa, Australia and the state of California. It is an excellent guide fossil. The example in the photograph, which belongs to the species *C. sandalina*, comes from the Devonian (380 million years ago) of Eifel, Germany, and measures 2 cm (¾ in).
Note This animal's shape is reminiscent of certain Brachiopoda and Bivalvia adapted to life in the reef, and for a long time this fossil was not regarded as a coral at all. *Calceola* probably lived with its ventral side (the convex one) resting on the seabed and was able to withdraw when necessary by closing its operculum.

27 MEANDRINA

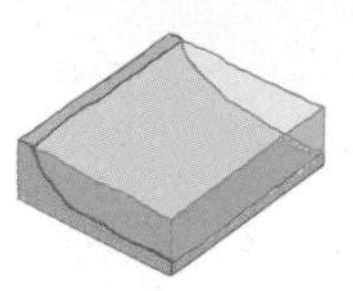

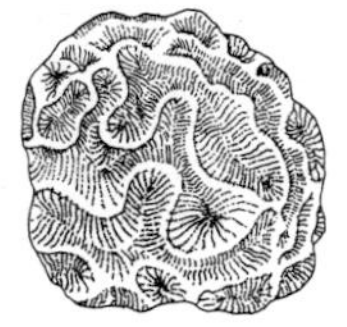

Classification Phylum Coelenterata, Class Anthozoa, Order Scleractinia, Family Meandrinidae.
Description The appearance of a colony of *Meandrina* is similar to a meandering maze, with the corallite fused laterally to create a series of winding, discontinuous grooves whose series are joined together by a wall (septotheca) unaffected by ambulacra. The columella is generally discontinuous; the septa are straight, laminar and positioned perpendicularly to the "hills" that separate the grooves. The genus *Meandrina* is also known as "brain coral" because its convolutions look like the cerebral matter in advanced vertebrates.
Stratigraphic position and geographical distribution The genus *Meandrina* dates in doubtful form from the Eocene and is still found in seas today. Its area of distribution comprises Europe, the West Indies and Brazil. The colony in the photograph measures 7 cm (2¾ in) and comes from the French Pliocene (4 million years ago).
Note The genus *Meandrina* today is found in the Atlantic, but forms related to *Meandrina* and belonging to the same family are also found in the Pacific.

---

28 FLABELLUM

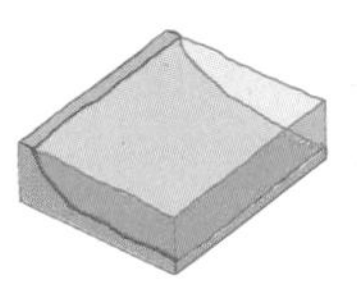

Classification Phylum Coelenterata, Class Anthozoa, Order Scleractinia, Family Flabellidae.
Description A solitary coral that can reach 10 cm (4 in) in height and possesses a wedge or circular fan shape (flabelliform). Its lifestyle is distinctive in that it lives unattached on the sea floor. The calice is deep and compressed, and characterized by the presence of numerous septa joined together at the center to form a false columella. There are no pores or dissepiments. Externally, the wall displays growth lines that tend to be superimposed.
Stratigraphic position and geographical distribution The genus *Flabellum* is known from the Eocene to the present day in sediments throughout the world. The example in the photograph comes from the Pliocene clays (7 million years ago) of Castell'Arquato, Piacenza (Italy).
Note This coral lives at depths ranging from 3 meters (10 ft) to 3,183 meters (9,840 ft). Its fossil remains occur particularly in Miocene and Pliocene deposits in the Tertiary Piedmontese basin and the Tuscan Apennines of Italy. It is now often found in the Azores (species *F. alabastrum*) and the Hawaiian Islands (species *F. pavonimum*).

29 HALYSITES

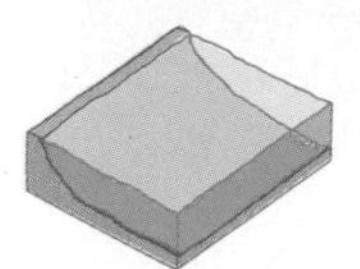

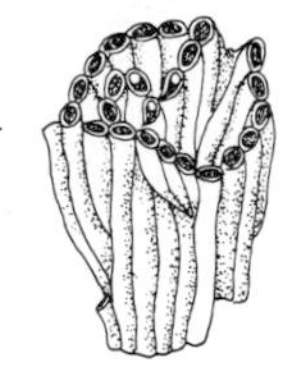

Classification Phylum Coelenterata, Class Anthozoa, Order Tabulata, Family Halysitidae.
Description A colony is composed of elongated corals, either cylindrical or oval in shape, joined together in a long chain that intertwines to form an irregular mesh. The general form of this chain recalls a rosary. The corallites are separated from each other by a microcorallite almost polygonal in form. In the genus *Halysites* the wall is imperforate and the calices are either totally lacking in septa or characterized by the presence of 12 vertically placed spines; the tabulae are well developed and either horizontal or gently arching.
Stratigraphic position and geographical distribution The genus *Halysites* is known from the Ordovician and Silurian. It has been identified in North America, Europe, Asia, Africa and Australia. The example in the photograph, belonging to the type species *H. catenularia*, comes from the English Silurian (400 million years ago). The entire colony measures 10 cm (4 in).
Note The family Halysitidae is represented by four genera that are distinguished from *Halysites* by the size of their tubes and the form of their corallites. Their stratigraphic position coincides with that of the genus *Halysites*, while the geographical distribution of the individual genera is extremely variable.

---

30 FAVOSITES

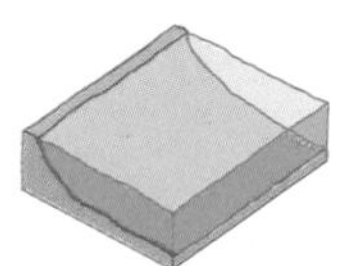

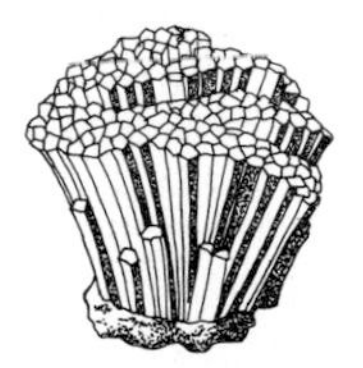

Classification Phylum Coelenterata, Class Anthozoa, Order Tabulata, Family Favositidae.
Description A colonial coral with a honeycomb appearance; the colony is massive and can reach considerable dimensions. The individual corallites have a prismatic cross section and are edged by a thin, porous wall. The calices may be characterized by the presence of septa, sometimes replaced by septal spines; these spines and septa can vary in number, but they are not much developed in any case. The coral possesses complete tabulae.
Stratigraphic position and geographical distribution The genus *Favosites* is characteristic of deposits from the Upper Ordovician to the Middle Devonian. It is found on all continents. The example in the photograph, which measures 12 cm (5 in), comes from the Upper Silurian (400 million years ago) on the Swedish island of Gotland.
Note The family Favositidae includes some of the most abundant forms of coral from the Silurian and Devonian periods; particularly common during these periods were members of the genus *Favosites*.

31 PLEURODICTYUM

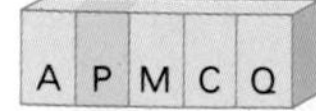

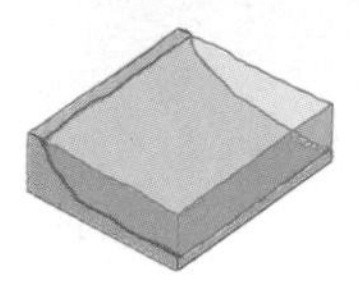

Classification Phylum Coelenterata, Class Anthozoa, Order Tabulata, Family Favositidae.
Description The colony is discoidal or hemispherical in form, with broad, polygonal corallites. The wall of the latter is thick and reveals irregularly distributed pores. It is also possible to detect septal spines and complete, relatively thin tabulae. The colony generally establishes itself on brachiopod shells or worm tubes.
Stratigraphic position and geographical distribution The genus *Pleurodictyum* occurs in sediments of the Lower Devonian (380 million years ago) and is widely distributed. The example in the photograph, a member of the species *P. problematicum*, measures 3 cm (1¼ in) in diameter and comes from the Eifel in Germany.
Note The family Favositidae, to which the genus *Pleurodictyum* belongs, is characterized by massive forms, typically without a coenenchyme; the corallites are thin and there are mural pores and short, spiny septa present in many different manifestations. The family Favositidae is present from the Upper Ordovician to the Permian; attribution of these corals to the Triassic is highly debatable.

---

32 LEIOCHAETETES

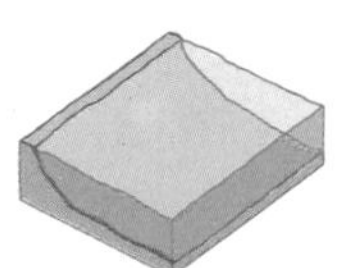

Classification Phylum Porifera, Class Sclerospongia, Order Chaetetida, Family Achanthochaetetidae.
Description An encrusting colony conical in appearance, bell-shaped and less than 2 mm (less than ¼ in) thick. The tubules are polygonal, of variable dimensions and placed one next to the other; their wall is very thin. There are tabulae, but these are scarce; they are thin and slightly convex. There are no noticeable spines in the lumen of the tubules. The colony increases basally and radially.
Stratigraphic position and geographical distribution The genus *Leiochaetetes* is known in the Upper Tithonian layers (Upper Jurassic, 120 million years ago) of Passo del Furlo in the Italian Apennines. The fossil in the photograph, whose diameter is approximately 10 cm (4 in), is an example from this region.
Note The systematic positioning of the Chaetetidae has been, and still is, open to discussion. They have been placed among Anthozoa, Hydrozoa and Bryozoa in the past. Their inclusion in the phylum Porifera is done with certain reservations.

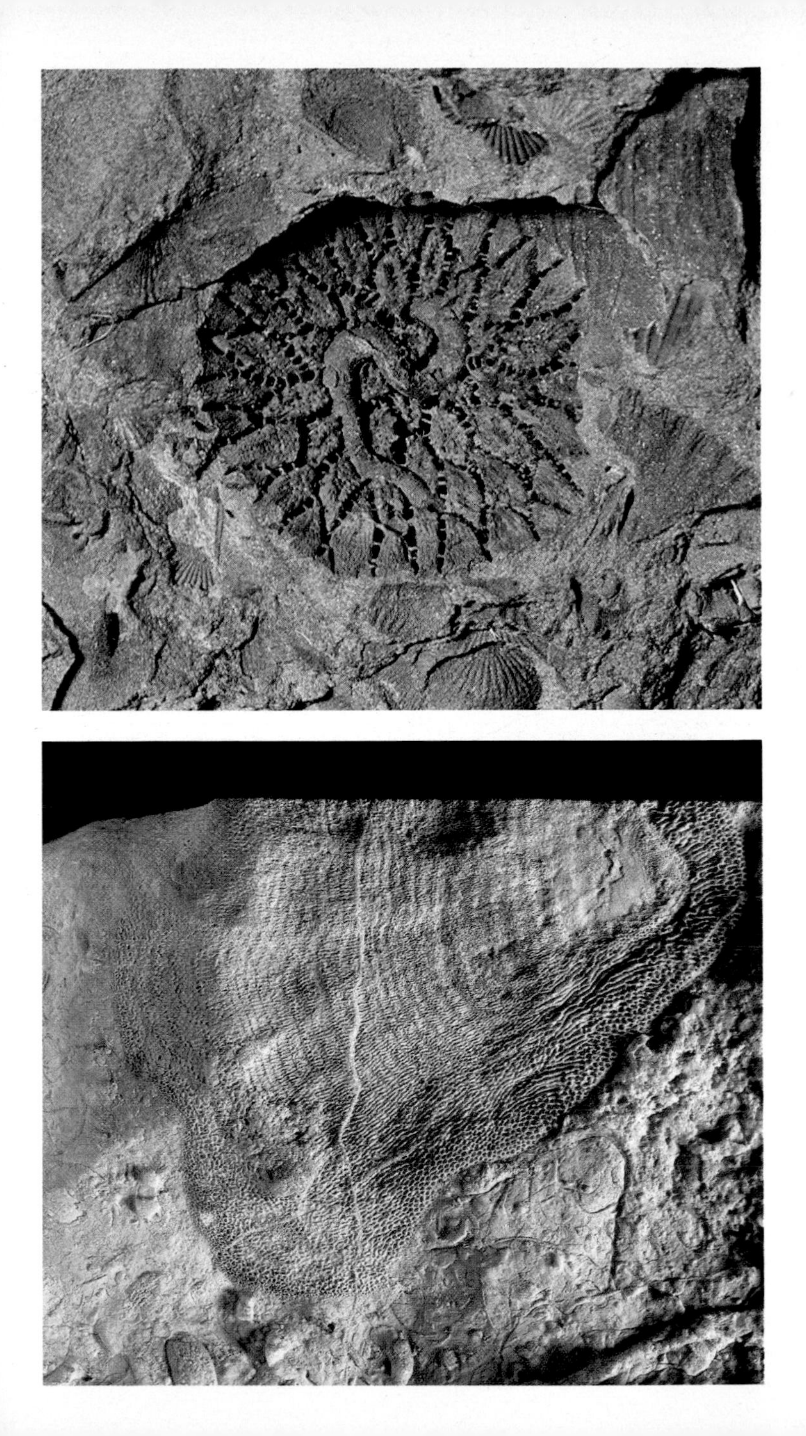

33 ARCHIMEDES

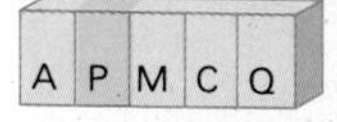

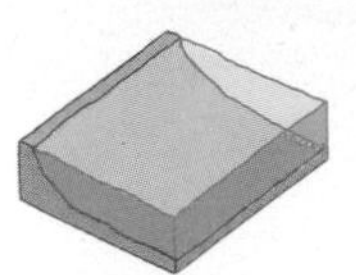

Classification Phylum Bryozoa, Class Gymnolaemata, Order Cryptostomata, Family Fenestellidae.
Description A colony that develops along an axis from which there emerges, in a corkscrew arrangement, a reticular netlike structure composed of bars running parallel to the axis toward the periphery and linked together by other structures positioned perpendicularly to the bars themselves. The apertures of the zooecia are situated at the upper end of these long bars, arranged in two rows and in different ways: either positioned opposite each other or staggered.
Stratigraphical position and geographical distribution The genus *Archimedes* appears in sediments dating from the Upper Carboniferous to the Permian, both in Europe and North America. The example in the photograph is from the Mississippian stratum (330 million years ago) of the states of Indiana, Iowa and Illinois, and measures approximately 7 cm (2¾ in).
Note The order Cryptostomata is characterized by Bryozoa possessing a very delicate calcareous lattice structure. The zooecia bear an operculum in the region of the mouth. The order Cryptostomata is known through fossil discoveries in sediments ranging from the Ordovician to the Permian.

---

34 LINGULA

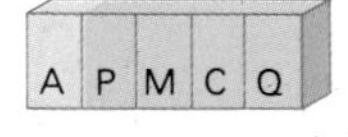

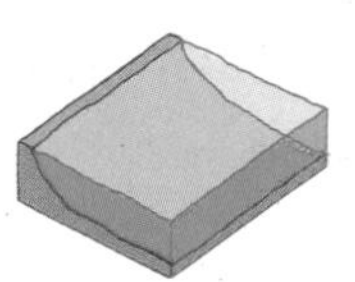

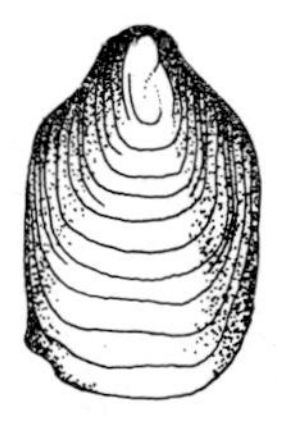

Classification Phylum Brachiopoda, Class Inarticulata, Order Lingulida, Family Lingulidae.
Description The shell is in the form of an elongated oval, with the lateral edges slightly convex or subparallel, and its outer surface is characterized by concentric growth lines. It is thin and possesses an internal cavity bereft of septa. There are numerous muscular impressions on each valve, but they are hard to detect; close to the anterior end of both valves there are traces of two distinct pallial sinuses.
Stratigraphic position and geographical distribution The genus *Lingula* has been found in rocks dating from the Ordovician, but doubts still remain concerning these discoveries; it is, however, clearly present from the beginning of the Silurian and is still found widely distributed in coastal waters today. The example in the photograph, which belongs to the species *L. cuneata*, comes from the Silurian (400 million years ago) in the state of New York; it measures 1 cm (½ in).
Note The shells of animals belonging to the order Lingulida are, with one exception, formed of alternate lamellae of chitin and calcium carbonate. The valves are biconvex and possess a pedicle that emerges between the two valves at the posterior end. Modern Lingulida burrow in sediment, as did the fossil forms. The habitat of these brachiopods is strictly marine, although some of them can adapt themselves to life in waters with a reduced level of salinity.

35 DISCINISCA

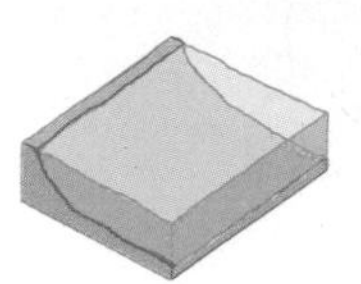

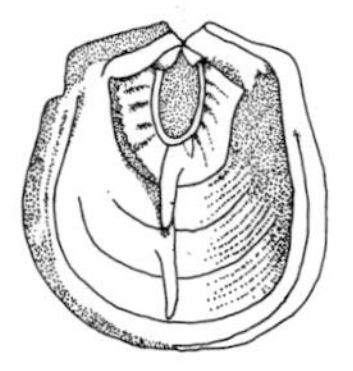

Classification Phylum Brachiopoda, Class Inarticulata, Order Acrotretida, Family Discinidae.
Description The shell is irregularly subcircular in form, ranging from biconvex to concave-convex, with the apex situated slightly off center. Its external ornament is provided by concentric growth lines; growth occurs multilaterally, in very thin costae. The interior of the animal is characterized by a highly developed musculature consisting of two pairs of muscles, one anterior and one posterior, which serve to close the valves of the shell, and a series of oblique muscles that open them. The pedicle valve contains a subtriangular median septum. The animal anchors itself by means of the pedicle emerging from the pedicle valve.
Stratigraphic position and geographical distribution The family Discinidae ranges from the Ordovician (500 million years ago) to the present day in forms that are in some cases widespread and in others confined to North America, Europe and Asia. The genus *Discina* is still found in the Atlantic, along the coastline of West Africa. The example in the photograph, of the species *D. calymene*, comes from the Lower Triassic (Scythian stage, 245 million years ago) at Val Badia (Italy).
Note The shell of this inarticulate brachiopod is made of calcium phosphate. The example in the photograph measures approximately 1.5 cm (1½–1¾ in).

---

36 STROPHOMENA

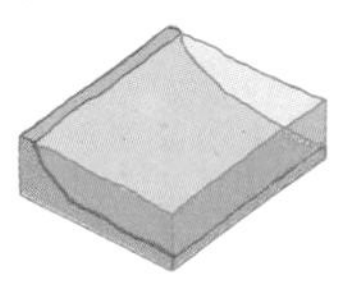

Classification Phylum Brachiopoda, Class Articulata, Order Strophomenida, Family Strophomenidae.
Description The shell ranges from small to medium-sized, is spherical in shape and has strongly biconvex valves. The anterior portion of the shell is plicate; in the dorsal region there is a strong fold, matched, in the ventral region, by an equally clear sulcus. The external ornament of the shell is provided by thin radial striae and inside there are very strong dental plates with a subparallel alignment. The cardinal process is small and knobbly. The very low notothyrial platform carries strong, well-developed, tooth-shaped brachiophores, which emerge from divergent bases.
Stratigraphic position and geographical distribution The genus *Strophomena* appears in the Middle Ordovician and becomes extinct in the Upper Ordovician; it possesses numerous species throughout the world. The example in the photograph, which measures approximately 2 cm (¾ in), comes from Upper Ordovician rocks (440 million years ago) in the state of Ohio.
Note The suborder Strophomenidina contains brachiopods whose shells vary from biconvex to concave-convex and which are characterized by a generally well-developed pseudodeltidium and chilidial plate. The hinge tooth is simple and only rarely either accompanied by additional teeth or denticles, or replaced by dental plates. The cardinal process is commonly bilobate. This suborder, with Paleozoic associations, became extinct in the Triassic.

37 LEPTAENA

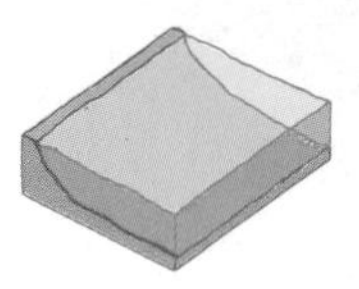

Classification Phylum Brachiopoda, Class Articulata, Order Strophomenida, Family Leptaenidae.
Description The shell appears flat-convex when the specimen is in its juvenile stage and becomes concave-convex with maturity. The outer surface is characterized by large numbers of concentric corrugations. The valves are characterized by a row, sometimes radially arranged, of tubular striae, which in well-preserved specimens are crossed by thin concentric striae. The cardinal margin is straight and as long as the maximum diameter of the shell. The interarea is narrow, slightly broader in the pedicle valve and has no denticulation. In the pedicle valve the delthyrium is covered by a convex deltidial plate that is perforated round the top. The teeth are very divergent and generally supported by laminae. The cardinal process consists of two divergent apophyses, with a striated attachment surface, which stretch along the edge. The indentations are moderately deep; the crural plates are not clearly defined.
Stratigraphic position and geographical distribution The genus *Leptaena* occurs in sediments from the Middle Ordovician to the Devonian and is widespread. The example in the photograph, measuring 2 cm (¾ in), comes from the Devonian (350 million years ago) of the Eifel in Germany.
Note The species *L. rhomboidalis* is an excellent guide fossil for the English Upper Silurian.

---

38 ATRYPA

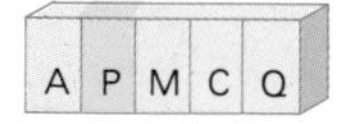

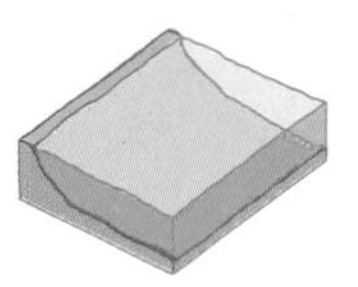

Classification Phylum Brachiopoda, Class Articulata, Order Spiriferida, Family Atrypidae.
Description A subcircular or longitudinally suboval convex shell, with markedly unequal valves. It has a short straight hinge line, rounded cardinal extremities and an inconspicuous beak. The pedicle valve is the smallest: convex in the umbonal region, depressed and sinuous in the anterior section. The beak, small and normally curving, conceals the foramen and the delthyrium. The foramen is triangular in juvenile specimens, extending along the cardinal margin, but is progressively inclosed by the growth of the deltidial plate until, at maturity, it becomes circular and apically positioned. The teeth are broad and separate. The convex brachial valve has a median fold solely near the anterior margin. The long narrow crura, are laterally divergent and placed on the primary layer. The spirals have bases parallel to the surface of the pedicle valve, with the tips pointing toward the deepest point on the valve opposite, whose axis is roughly convergent. The external surface has radial striae and concentric growth lines; sometimes there are spines.
Stratigraphic position and geographical distribution The genus occurs from the Lower Silurian to the Upper Devonian. The example *A. reticularis*, measures 1.5 cm (½–¾ in) and comes from the Middle Devonian (370 million years ago) of Westphalia (Germany).
Note The ripples along the margins of the valves of *Atrypa* have been interpreted, with certain reservations, as structures designed to support the shell in a muddy substratum.

39 ATHYRIS

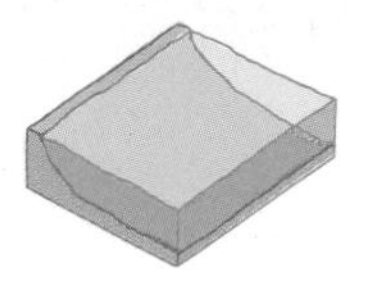

Classification Phylum Brachiopoda, Class Articulata, Order Spiriferida, Family Athyriidae.
Description The shell is biconvex, transversally elliptical, subcircular or suboval. The outer surface is moderately sinuate. In the pedicle valve the beak curves over to cover the foramen and deltidial plates. The valve is more convex in the umbonal region, whence descend, uniformly inclined in both directions, two layers of shell giving rise to a median depression; this widens out into a sinus that reaches its greatest extent in the region of the anterior margin. The beak on the brachial valve is not prominent. Within the pedicle valve the deltidial plates are generally absent, but teeth are evident, curved at the tip and supported by dental laminae. The dental fossae are broad and deep in the brachial valve. The brachidium consists of spiral cones placed base to base, with their tips pointing laterally. The shape of these cones varies with the internal cavity; they are normally vertically compressed. The long crura are convergent. The surface of the valves is decorated; a characteristic feature of this genus are the concentric growth lines bearing lamellae; the surface may be either smooth or with concentric striae.
Stratigraphic position and geographical distribution The genus *Athyris* ranges from the Lower Devonian to the Triassic, worldwide. The example, 1.5 cm (½–¾ in) in diameter, is from the Middle Devonian of the state of New York.
Note Representatives of the family Athyriidae make good guide fossils and are widely used for correlative research.

---

40 SPIRIFER

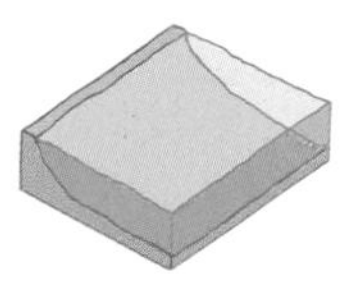

Classification Phylum Brachiopoda, Class Articulata, Order Spiriferida, Family Spiriferidae.
Description The shell is transversally elongate and can sometimes have a median fold and a sulcus. The hinge line is straight and usually represents the greatest diameter of the shell, which can be alate, sharp or rounded. The outer surface displays variously grouped granulations, striae, plicae and costae, which may be absent at the point where the fold and the sulcus occur. The striations are crossed by concentric growth lines that take the form of lamellae or tend to form spines. The pedicle valve has an umbo that is prominent in relation to the hinge line, with an acute, erect or curving apex. The cardinal area is flat or slightly curved, with transversal striae. Articulation is achieved by means of a tooth placed at the marginal extremities of the delthyrium, which is triangular; the dental plates are short. The brachial valve displays no marked umbo, while the cardinal area is narrow and divided by the delthyrium. The dental fossae are shallow and narrow; the cardinal process is transversal and the crura are long, narrow and divergent.
Stratigraphic position and geographical distribution The genus *Spirifer* occurs worldwide. Its stratigraphic distribution ranges from the Devonian to the Permian (390–240 million years ago). The example, *S. duodenarius*, comes from the Devonian (380 million years ago) at Sylvania, Ohio; it measures 3 cm (1¼ in) in diameter.
Note *Spirifer* is a guide fossil for the Carboniferous.

41 RHYNCHONELLA

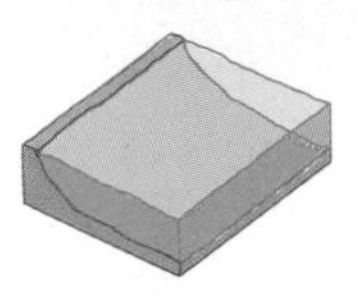

Classification Phylum Brachiopoda, Class Articulata, Order Rhynchonellida, Family Rhynchonellidae.
Description The shell is subpyramidal, the valve margins sinuate or angulate. The pedicle valve has a median sulcus that begins at the convex umbo and continues, broad and deep, to produce a tongue-shaped extension in the anterior margin. The brachial valve is convex in the umbonal region and develops anteriorly with a prominent median fold. The surface of both valves is more or less plicate, sometimes with a decoration of fine, concentric lines. The tip of the pedicle valve is slightly curved with a circular or elongated foramen, enclosed from below by the deltidial plates and from above by the valve. This forms a narrow area bounded by the oblique cardinal crests diverging from the beak. Inside, the teeth are well developed and supported by lamellae. There is no cardinal process in the brachial valve; the crural plates are simple and divergent, somewhat expanded, but never joined. The long crura curve upwards toward the valve opposite. The elongated muscular area has small posterior muscles and a large anterior adductor muscle.
Stratigraphic position and geographical distribution The genus *Rhynchonella* is characteristic of the European Jurassic. The example in the photograph comes from the Lias (160 million years ago) at Gozzano (Italy); it measures 2.5 cm (1 in).
Note *Rhynchonella* was a typical inhabitant of the so-called "marginal" marine area of the carbonate platform; it is therefore a good environmental indicator.

---

42 TEREBRATULA

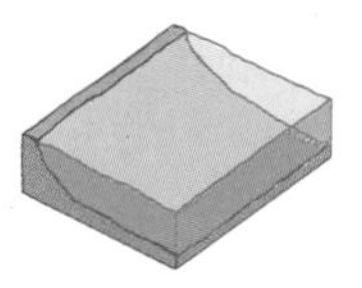

Classification Phylum Brachiopoda, Class Articulata, Order Terebratulida, Family Terebratulidae.
Description Shell of dimensions ranging from medium to large, with valves that are biconvex and anteriorly biplicate. The front commissure is uniplicate or sulcate; the umbo is short, massive and ranges from almost erect to curved. The foramen possesses a well-developed pedicle collar. The outer surface of the shell is smooth and shows growth lines that are sometimes very prominent. The loop is markedly triangular, stretching for approximately a third of the length of the valves; there are also ribbons. The cardinal process is rounded; the hinge plates are concave and separated by a deep sulcus. The hinge tooth has a swollen base and is posteriorly sulcate.
Stratigraphic position and geographical distribution The genus *Terebratula* is known in sediments of the European Miocene and Pliocene. The example in the photograph, *T. ampulla*, is 2.5 cm (1 in) long and comes from the Pliocene clays (5 million years ago) at Asti (Italy).
Note The genus *Terebratula* belongs to the suborder Terebratulidina, which survives today. Representatives of this suborder appeared in the Lower Devonian; during the Paleozoic only a few genera can be assigned to this group. In the Mesozoic these animals spread rapidly, but they were reduced to a few genera during the Cenozoic. The genus *Terebratula* occurs abundantly in the Pliocene sediments of the Piedmontese Tertiary basin.

43 TENTACULITES

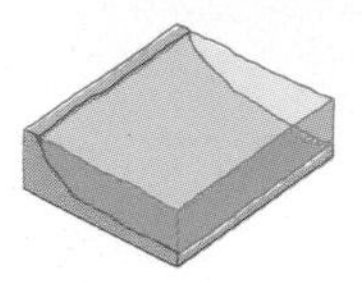

Classification Phylum Mollusca, Class Tentaculita.
Description An elongate, almost needle-shaped calcareous shell, straight, closed and pointed at one end. The exterior of the shell displays a decoration of ribbed rings. Exclusively marine organisms, tentaculitids are regarded as benthonic forms, meaning ones living on the seabed in shallow waters.
Stratigraphic position and geographical distribution *Tentaculites*, known from the start of the Middle Cambrian (500 million years ago), can be traced up to the Devonian (345 million years ago). Their area of distribution is restricted to Europe and North America. The example in the photograph, which belongs to the species *T. gyracanthus*, comes from the Silurian (420 million years ago) at Ravena, New York; it measures 2 cm (¾ in).
Note Tentaculitids are a class with a very uncertain taxonomic position: they have been attributed to a variety of phyla, but are generally regarded as a type of Mollusca because of their supposed similarities with the shells of certain modern Opistobranchia. Recently, however, studies of examples retaining their soft parts have shown that certain tentaculitids had an organization similar to that of mollusc Cephalopoda, since they possessed a siphuncle and tentacles, whereas other examples display a shell structure similar to that of Brachiopoda. In some Silurian and Devonian rocks the remains of tentaculitids are so abundant that they represent one of the main constituent elements.

---

44 CONULARIA

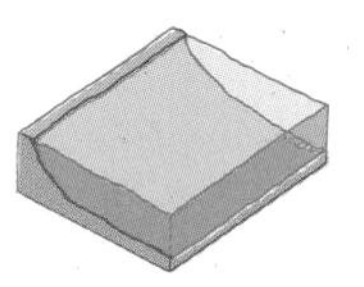

Classification Phylum Conulata.
Description These enigmatic fossils, whose zoological affinities with other animal phyla are unknown, consist of the remains of very thin tests or skeletons composed of chitin and some calcium phosphate. The skeletons have a characteristically pyramidal form. They display fine, clearly visible striations running across the length of the shell, which in all probability correspond to the animal's growth lines. In section, the shell appears square or rhomboidal and there are also internal septa.
Stratigraphic position and geographical distribution The earliest conulariids, which are exclusively marine fossils, are known from the Cambrian; sometimes very frequent in Paleozoic rocks, they went into rapid decline at the end of that era, finally disappearing in the Triassic period at the beginning of the Mesozoic era. The example in the photograph, which belongs to the species *C. crustula*, comes from the Pennsylvanian stratum (Upper Carboniferous, 300 million years ago) at Wyandotte, Kansas; it measures 2 cm (¾ in).
Note Classified in the past as scyphozoan coelenterates, making them akin to jellyfish, Conulariida were later recognized as a separate group. The organization of their bodies is still unknown, but it does appear that some forms may have lived attached to floating objects, while others were either carried by the currents or possessed a capacity for autonomous swimming.

45 HYOLITHES

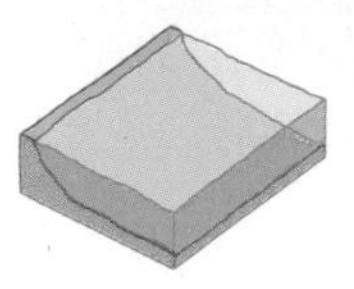

Classification Phylum Mollusca, Class Hyolitha.
Description *Hyolithes* is the most typical genus of the class Hyolitha, tentatively attributed to the phylum Mollusca. It possesses a conical shell, pointed at the apex and equipped with an operculum at the opposite end; it also displays bilateral symmetry and appears roughly triangular in transversal section. There is a distinctive semicircular expansion before the aperture. A pair of curving calcareous appendages project outside the shell, emerging from between the operculum and the shell itself, as can be seen in certain particularly well-preserved specimens. Given the thickness of the shell and its sometimes considerable size (up to 15 cm or 6 in), it is thought that *Hyolithes* was incapable of swimming and must therefore have lived on the sea floor without being able to indulge in active movement.
Stratigraphic position and geographical distribution Hyolithids were very abundant in the Cambrian, after which they went into a slow decline and finally became extinct in the Permian. The example in the photograph, belonging to the species *H. cecrops*, comes from the Middle Cambrian (500 million years ago) at Liberty, Idaho; it measures 2.5 cm (1 in).
Note The classification of hyolithids poses considerable difficulties: in the past they were regarded as Gastropoda, or lumped together with Conulariida, even though the latter's skeletal composition is very different. Some paleontologists still regard them as forming a phylum of their own.

---

46 DENTALIUM

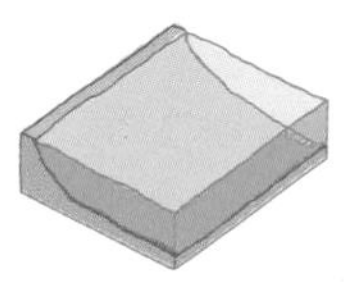

Classification Phylum Mollusca, Class Scaphopoda, Family Dentaliidae.
Description The shell of the genus *Dentalium* is tubular in appearance, broader in the anterior section and gradually tapering posteriorly; it is elongate and bilaterally symmetrical. Its composition is calcareous and there are openings at both ends of the curved shell. The dorsal region is located on the inside of the curve, the ventral region on the outside. The mouth is located at the anterior. The animal crawls by moving a foot that is more or less cylindrical in shape. The ornament on the shell is provided by longitudinal ribbing accompanied by fine granulation.
Stratigraphic position and geographical distribution The genus *Dentalium* is first found in strata of the Middle Triassic (220 million years ago) and is still living today; it occurs throughout the world. The example in the photograph, which belongs to the species *D. caloosaensis*, comes from Pliocene sediments (5 million years ago) in the state of Florida; it measures approximately 6 cm (2½ in).
Note This strange mollusc lives with the anterior region of its shell imbedded in the sea floor. It is exclusively marine and can live at great depths. It feeds on small organisms, such as benthonic Foraminifera, which it captures by means of the large numbers of small tentacles that adorn its mouth.

47 PTEROCHITON

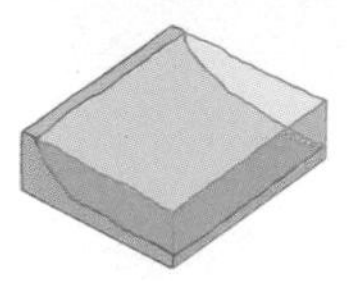

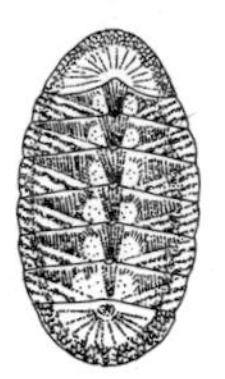

Classification Phylum Mollusca, Class Amphineura, Order Neoloricata, Family Lepidopleuridae.
Description A marine mollusc belonging to the subclass Polyplacophora, bilaterally symmetrical. It is characterized by a series of eight articulated calcareous plates; the anterior and posterior plates are rounded, while the intermediate ones are rectangular, with a spiny process behind. The animal has no eyes. In the area surrounding the plates (the girdle) there may be granules, scales, calcareous spines and spiny fringes. If disturbed, it has the ability to roll itself into a ball.
Stratigraphic position and geographical distribution The genus *Pterochiton* is found in Lower Carboniferous strata (340 million years ago) in Belgium and Ireland, whereas in Mazon Creek, Illinois, it occurs in sediments of the Upper Carboniferous (290 million years ago). The example in the photograph, belonging to the species *P. concinnus*, comes from the state of Illinois.
Note Polyplacophora are animals adapted to a herbivorous diet, but there are also omnivorous forms. They are animals that live on the sea floor, sometimes at great depths.

---

48 GLYCYMERIS

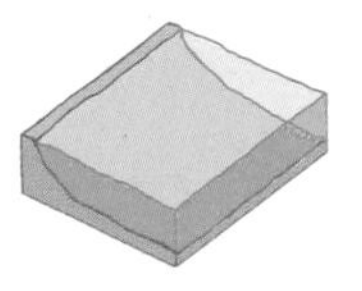

Classification Phylum Mollusca, Class Bivalvia, Order Arcoida, Family Glycymerididae.
Description The shell of *Glycymeris* is orbicular in shape with a fairly strong, almost equilateral structure. The valves are rather swollen and possess faintly pronounced umbones. The ligament is external and the ligament area is broad, triangular in shape and marked with striations. The hinge is composed of numerous teeth, which become more or less slanting as they diverge from the center of the hinge. Inside the shell there are two muscle scars, subequal and roundish, while the pallial line has no pallial sinus. The external ornament is characterized by radial costae and faint, concentric growth lines.
Stratigraphic position and geographical distribution The genus *Glycymeris* appears in strata of the Lower Cretaceous (130 million years ago) and still occurs today. It is found throughout the world. The example in the photograph, which belongs to the species *G. parilis*, comes from Pleistocene layers in the state of Florida and has a diameter of 4.5 cm (1½ in).
Note *Glycymeris* is a free-moving inhabitant of sandy coastal areas and is a particularly active animal.

49 PINNA

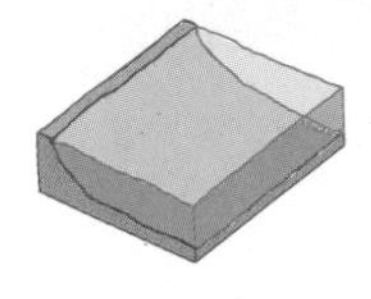

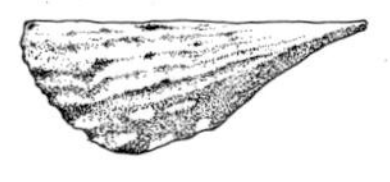

Classification Phylum Mollusca, Class Bivalvia, Order Mytiloida, Family Pinnidae.
Description The shell of *Pinna* is triangular, almost equivalve and drastically inequilateral; it is anteriorly pointed, while enlarged and rounded at its posterior. Inside, the nacreous layer is subdivided by a median keel that stretches the full length of the valve; there is also a clearly visible muscle scar. The hinge is very much reduced and dysodont, whereas there is a highly developed external ligament. The external ornament is composed of a straight, prominent median ridge and radial costae that tend to bear thin lamellae in the ventral region.
Stratigraphic position and geographical distribution The genus *Pinna* is known from the Lower Carboniferous (340 million years ago) and still occurs in seas today; as a fossil it is found throughout the world. The example in the photograph was discovered in Upper Triassic strata (200 million years ago) in Lombardy (Italy); it measures 5 cm (2 in).
Note The shell of *Pinna* is characterized in its adult stage by a byssus with which it affixes itself to the sea floor or to rocks.

---

50 INOCERAMUS

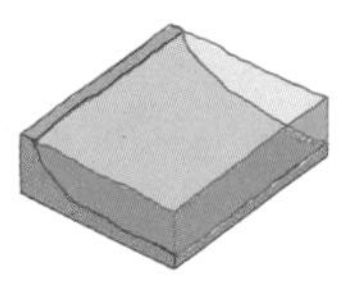

Classification Phylum Mollusca, Class Bivalvia, Order Pterioida, Family Inoceramidae.
Description The shell of *Inoceramus* is characterized by a shape that ranges from the subquadrangular to the oval, normally taller than it is long. The left valve is generally more convex than the right one. The hinge line is straight and there is a multivincular ligament positioned over a series of small parallel grooves. The interior reveals a clearly visible, continuous pallial line. The musculature is formed of two adductor muscles, one of which is distinctly more developed. The external ornament is provided by numerous concentric plicae.
Stratigraphic position and geographical distribution The genus *Inoceramus* appears in the Lias (Lower Jurassic, 190 million years ago) and disappears in the Upper Cretaceous (80 million years ago). Representatives of this genus have been discovered in Europe and also on other continents. The example in the photograph, which belongs to the species *I. balticus*, comes from Cretaceous strata in Germany and measures 7 cm (2¾ in) in diameter.
Note *Inoceramus* was a benthonic form that affixed itself by means of a byssus or by cementing the right valve to the rigid substratum. It is a good guide fossil.

51 ISOGNOMON

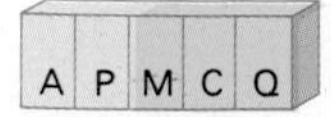

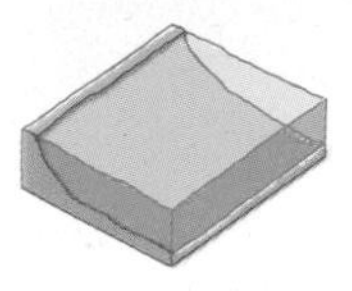

Classification Phylum Mollusca, Class Bivalvia, Order Pterioida, Family Isognomonidae.
Description The shell of *Isognomon* is subequivalved, with a height usually greater than its length, and has a subquadrangular outline. The shape is rather compressed and reveals anteriorly facing and curving umbones. The ligament is multivincular and formed of a series of rounded crests regularly interspersed with small grooves; the crests and grooves are straight and parallel, they run transversally across the whole margin and are sited on top of a flat area. Inside, there is a single muscle scar and a discontinuous pallial line. The external ornament is characterized by thin, concentric growth lines that may be accompanied by lamellae. There are no radial ornamental elements.
Stratigraphic position and geographical distribution The genus *Isognomon* is present from the start of the Upper Triassic (210 million years ago) and survives today; it is a widespread genus. The photograph shows a representative of the species *I. maxillatus*, measuring 12 cm (5 in), from the Pliocene at Asti (Italy).
Note Shells of *Isognomon* discovered in Pliocene sediments are very hard to preserve intact because they tend to flake progressively in layers perpendicular to the ligament area.

---

52 AVICULOPECTEN

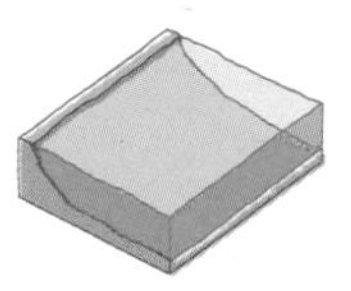

Classification Phylum Mollusca, Class Bivalvia, Order Pterioida, Family Aviculopectinidae.
Description The shell of *Aviculopecten* is oval in appearance, with gently convex valves characterized by two small, almost equal "ears" (although the posterior "ear" can be longer) placed in the umbonal region. The umbones are shallow. Among the characteristic features to be seen inside the shell is the resilifer, which is unique in its arrangement: it is, in fact, triangular, sited at a slant almost at the center of the hinge. The ornament of the shell is provided by concentric growth lines and irregular, radially-arranged costae. The byssus emerges from the right valve.
Stratigraphic position and geographical distribution The genus *Aviculopecten* is found in the Lower Carboniferous strata (Mississippian, 345 million years ago) and became extinct in the Upper Carboniferous (Pennsylvanian, 280 million years ago). The example in the photograph, which belongs to the species *A. mazonensis*, comes from the Upper Carboniferous of Mazon Creek, Illinois; it measures 3 cm (1¼ in).
Note *Aviculopecten* occurs with a certain degree of frequency in the Mazon Creek beds. This bivalve lived near the coastline and attached itself byssally to the sea floor.

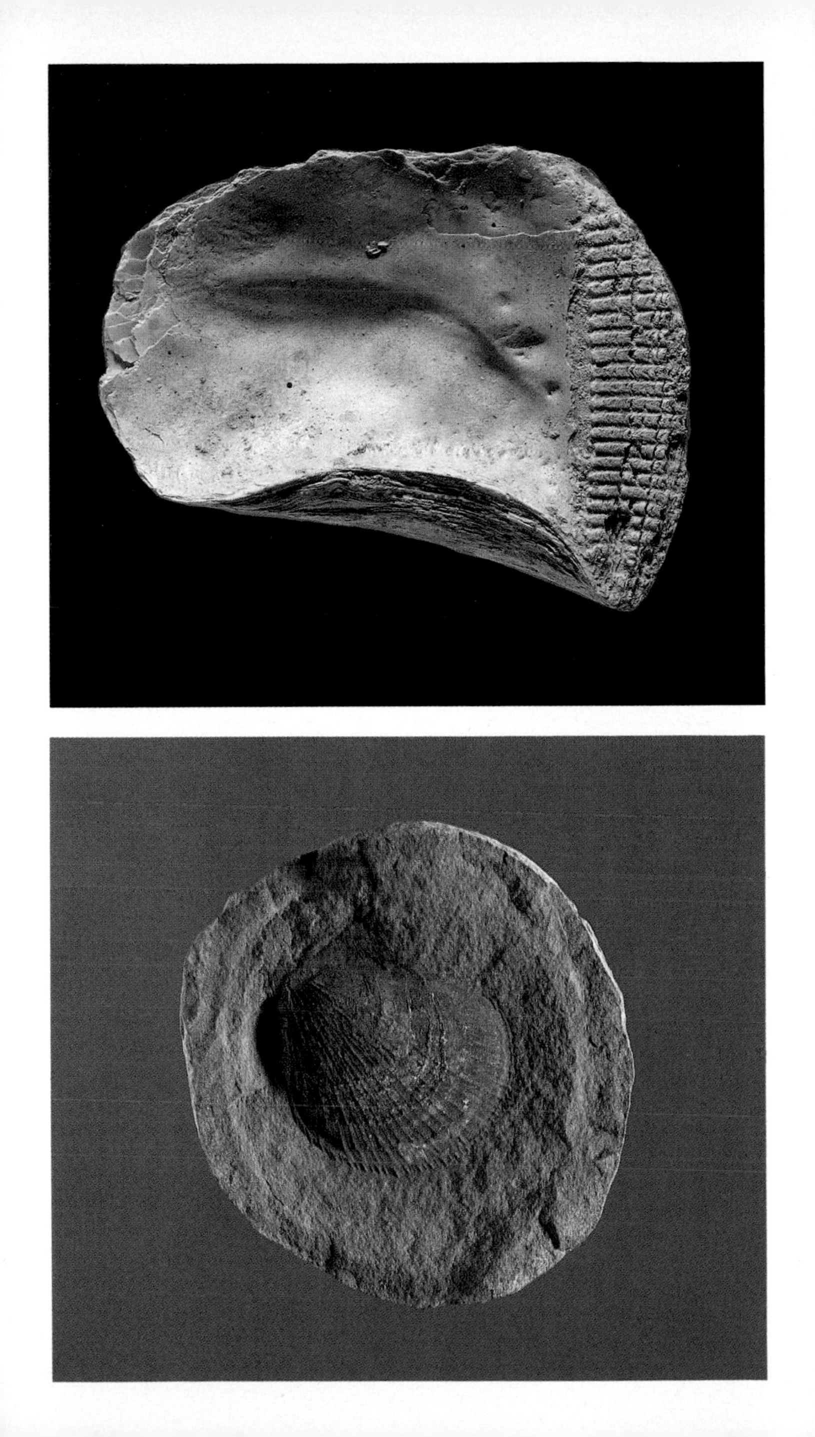

53 POSIDONIA

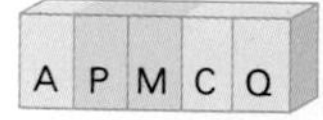

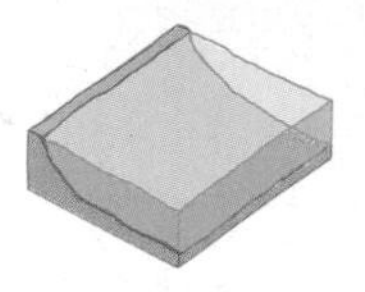

Classification Phylum Mollusca, Class Bivalvia, Order Pterioida, Family Posidoniidae.

Description The shell of the genus *Posidonia*, which is extremely thin, ranges in shape from oval to rhombic. Slightly inequivalved and inequilateral, it is gently concave and characterized by a short, straight dorsal margin with very small or even nonexistent "ears." The umbones are very poorly developed, the ligament area is triangular and the hinge lacks teeth. Inside, there is a curious muscle scar and an entire pallial line. The external ornament is provided by barely visible costae, radially arranged, and characteristic wavy concentric lines.

Stratigraphic position and geographical distribution The earliest known fossil form of *Posidonia* comes from sediments of the Lower Carboniferous (340 million years ago), but the genus became extinct at the end of the Cretaceous (65 million years ago). It is found in North America, South America, eastern Asia and Europe. The example in the photograph, which belongs to the species *P. bekeri* and measures 3.5 cm (1¼–1½ in), comes from the German Carboniferous.

Note *Posidonia* attached itself, by means of a byssus that emerged from the right valve, to floating logs and the shells of other molluscs or cemented itself directly to the sea floor by means of the right valve. It preferred calm, warm, oxygenated waters.

---

54 PECTEN

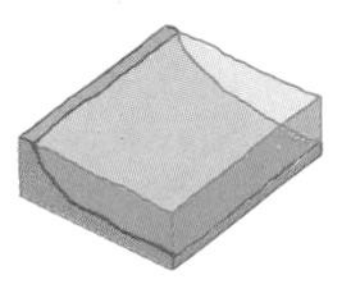

Classification Phylum Mollusca, Class Bivalvia, Order Pterioida, Family Pectinidae.

Description The shell of *Pecten* is subcircular and characterized by a convex right valve and a flat, slightly concave or convex left valve. *Pecten* possesses two "ears," identical in size and shape, on both valves; these "ears" accompany the hinge. The resilifer is triangular, sited in the central region of the hinge area and is accompanied by well-developed crura that emerge from the resilifer in opposite directions and on the same plane. The interior displays the adductor muscle scar near the center of the shell. The ornament of this well-known shell is composed of costae radiating uniformly from the beak.

Stratigraphic position and geographical distribution The genus *Pecten* first appears in the Upper Eocene (40 million years) and is still present in seas throughout the world. The example in the photograph, which belongs to the species *P. (Flabellipecten) flabelliformis* and measures 8 cm (3¼ in) in diameter, comes from Pliocene strata near Asti (Italy). The stratigraphic position of this species ranges from the Lower Miocene (22.5 million years ago) to the Upper Pliocene (3 million years) in Eurasia, while it still lives in the seas off North, Central and South America.

Note The genus *Pecten* is characteristic of sandy areas in fairly deep waters.

55 SPONDYLUS

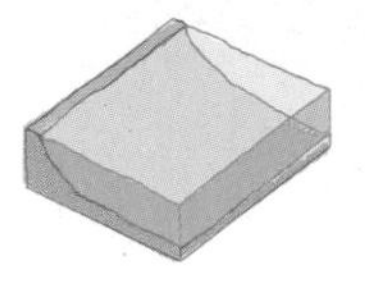

Classification Phylum Mollusca, Class Bivalvia, Order Pterioida, Family Spondylidae.
Description The shell of the genus *Spondylus* is orbicular, gibbous and equilateral in appearance. It is markedly inequivalved; in fact, it affixes itself to the sea floor by means of the right valve, which is more convex than the left one and possesses a highly developed umbo. The interior of the shell is characterized by the presence of a large muscle scar near the center; the hinge comprises two teeth placed on each valve. The pallial line is entire. The ornament is formed of radial costae the arrangement of which can be markedly irregular. In addition to this radial ornament, certain species possess more or less regular spines that may sometimes even be leaf-shaped.
Stratigraphic position and geographical distribution The genus *Spondylus* occurs throughout the world in fossil form in layers dating from the Lower Jurassic (190 million years ago) onward and is still widespread today. The photograph shows an example of the species *S. crassicosta* from the Pliocene at Asti (Italy); it measures 7 cm (2¾ in) in diameter.
Note This mollusc is characteristic of coastal environments with warm, oxygenated waters and rocky depths.

---

56 MYOPHORIA

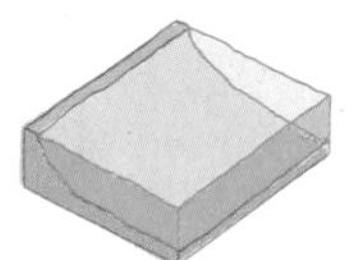

Classification Phylum Mollusca, Class Bivalvia, Order Trigonoida, Family Myophoriidae.
Description The shell of *Myophoria* is ovate-triangular in shape and dramatically inequilateral. It is characterized by the presence of a strongly defined, slanting keel marked by thin radial ribs (generally just one), which run for its entire length. This keel marks the limit of the posterior area, which is indented. The hinge is formed, on the left valve, by a very strong median tooth, which may be simple or forked, and by an underdeveloped posterior tooth. Internally there is a well-developed pad that accompanies the muscle scar. The external ornament of the shell is composed of concentric lines placed in front of the ridge and by a few radial costae.
Stratigraphic position and geographical distribution The genus *Myophoria* is a typical Triassic (230–200 million years ago) fossil form characteristic of European, Asian and North African sediments. The *Myophoria* fossil in the photograph, which measures 2.5 cm (1 in), comes from the Carnian sediments at Gorno in Italy.
Note In Italy Carnian sediments (210 million years ago) are the result of sedimentation occurring in a muddy, shallow water basin, with abundant supplies of terrigenous matter.

57 ARCINELLA

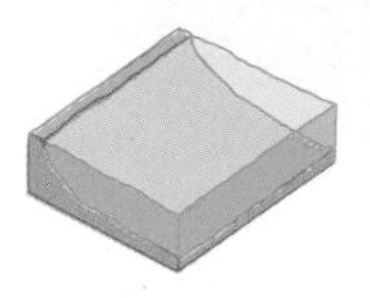

Classification Phylum Mollusca, Class Bivalvia, Order Veneroida, Family Chamidae.
Description The shell of the genus *Arcinella* is quadrangular in outline, with both valves strongly convex; the umbones are highly developed and prosogyrous. On the left valve is a hinge composed of two cardinal teeth; the right valve is characterized by the presence of a single cardinal tooth that separates two small fossae; also visible in the interior are two fairly broad and almost equal muscle scars; the pallial line is entire. The external ornament is provided by highly accentuated radial costae accompanied by large numbers of spines. In the early stages of growth the animal affixes itself to the sea floor or to the surface of a shell; in its adult stage it is free-moving.
Stratigraphic position and geographical distribution The genus *Arcinella* is known in fossil form from the Miocene (20 million years ago) and is still living today. Its geographical distribution, both as a fossil and as a living organism, is restricted to the U.S. and central-southern America. The example in the photograph, which measures 3 cm (1¼ in), comes from the Florida Pliocene and belongs to the species *A. cornuta*.
Note *Arcinella* is the sole genus of the family Chamidae to occur on the North American continent. Its natural habitats are sandy or muddy beds near the coast.

---

58 TRIGONIA

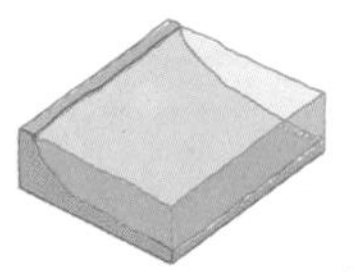

Classification Phylum Mollusca, Class Bivalvia, Order Trigonioida, Family Trigoniidae.
Description The shell of the genus *Trigonia* is generally triangular in shape, with convex valves the anterior side of which is rounded, the posterior side straight and truncate. A prominent keel, which emerges from the umbonal region, marks the posterior area, which is broad, flat and triangular. There is a second ridge, less prominent and shorter, bounding the escutcheon. The umbones are not prominent and are opisthogyrous. The hinge is formed on the left valve by a forked and indented central tooth and by two teeth, one positioned anteriorly and the other posteriorly, which are much less developed, particularly the posterior one. The right valve has two divergent, almost identical and near-symmetrical teeth. Inside the shell are two muscle scars. The external ornament on the anterior region differs from that on the posterior one: anteriorly there are fairly straight concentric lines, with rows of tubercles and costae, while posteriorly the concentric lamellae are decidedly thinner and denser.
Stratigraphic position and geographical distribution The genus *Trigonia* is known from the Middle Triassic (220 million years ago) and became extinct in the Upper Cretaceous (65 million years ago). It occurs in many different localities. The example in the photograph, of the species *T. navis*, comes from the German Middle Jurassic and measures 5 cm (2 in).
Note *Trigonia* inhabited the coastline. *Neotrigonia* is the sole family member still found in Australian waters.

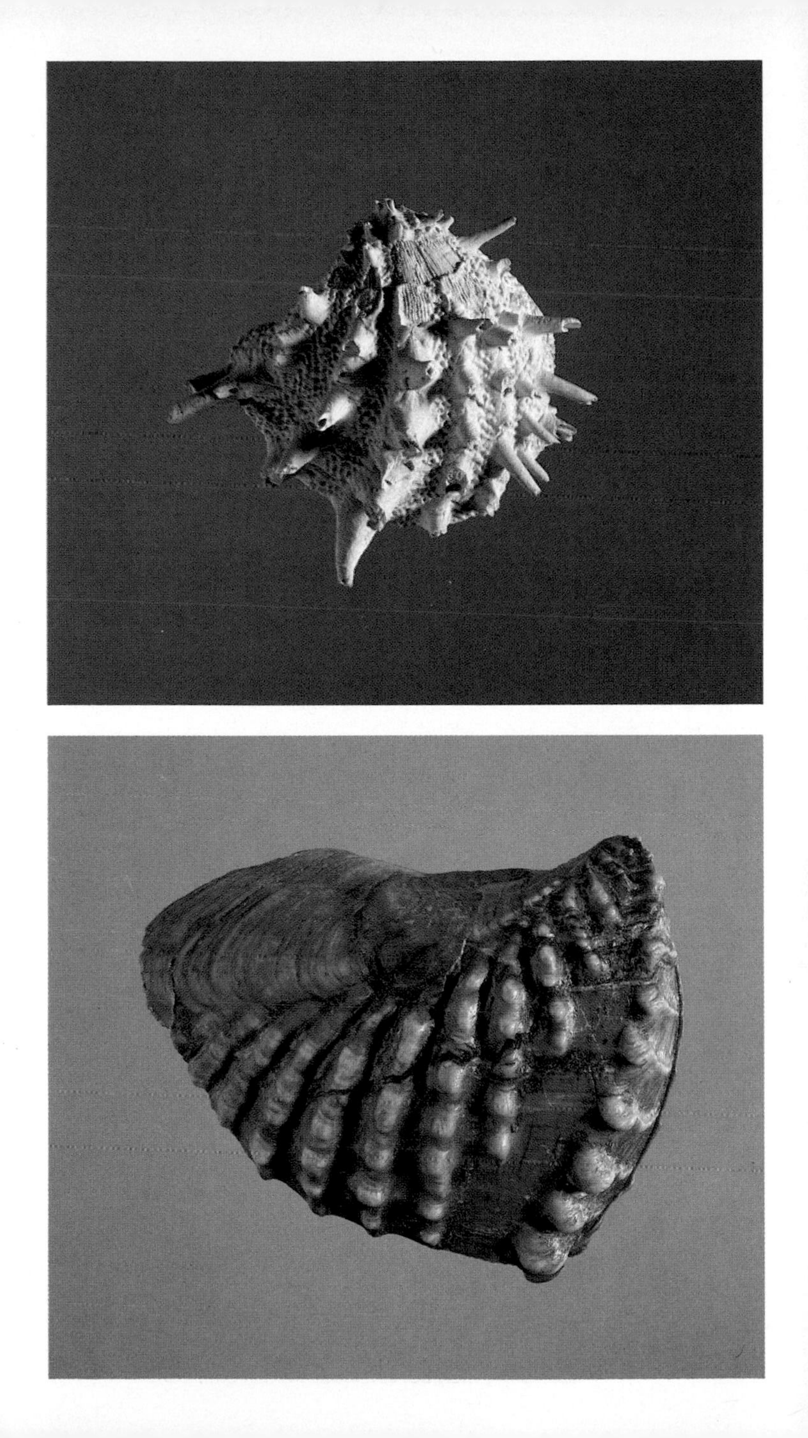

59 CARDIUM

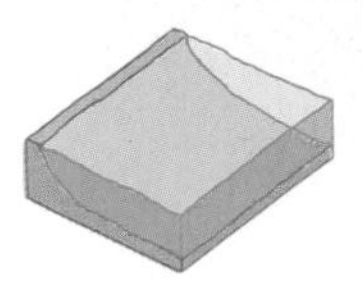

Classification Phylum Mollusca, Class Bivalvia, Order Veneroida, Family Cardiidae.
Description The shell of this bivalve is heart-shaped, with globular valves, and is more often than not greater in height than length; it is equivalved and inequilateral. The ligament is short and external. Inside, there are two conical hinge teeth of considerable height, positioned at right angles to the cardinal area on both valves. These hinge teeth are accompanied by an anterior lateral tooth and a posterior lateral one on the left valve, while the right valve may sometimes possess a second, anteriorly-placed lateral tooth, much reduced in size. The external ornament is provided by very prominent radial costae sometimes accompanied by weak spiny processes. Inside, there are clearly visible muscle scars, almost equal in size, and an entire pallial line.
Stratigraphic position and geographical distribution The genus *Cardium* appeared in the Miocene (20 million years ago), and is still present in seas today. Its geographical distribution is restricted to southern Europe and West Africa. The example in the photograph, which belongs to the species *C. ciliare*, comes from the Pliocene at Piacenza (Italy); it measures 3 cm (1¼ in) in diameter.
Note This is a typical bivalve that lives near the coastline in sandy or muddy sediments.

---

60 ARCTICA

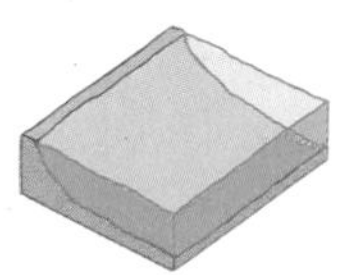

Classification Phylum Mollusca, Class Bivalvia, Order Veneroida, Family Arcticidae.
Description The shell of the genus *Arctica* is ovoid in shape and normally strong. It is equivalved and inequilateral, with no open regions. Internally it is characterized by a hinge composed of three cardinal teeth, present on each valve, accompanied by two anterior lateral teeth and a posterior tooth on the right valve and by only one anterior and posterior tooth on the left valve. The external ornament is provided by fine concentric growth lines accompanied by thin lamellae.
Stratigraphic position and geographical distribution The genus *Arctica* appears in the Lower Cretaceous (the Albian, 110 million years ago) and still occurs in seas today. It is found in fossil form in Europe and North America. The example in the photograph, a member of the species *A. islandica*, comes from the Sicilian Pleistocene and measures 10 cm (4 in).
Note This is a very important fossil from the point of view of paleontological reconstructions, since this animal, now a characteristic inhabitant of cold marine waters, infiltrated the Mediterranean basin during the cold phases of the Quaternary, thereby indicating that water temperatures dropped during that period.

61 VENUS

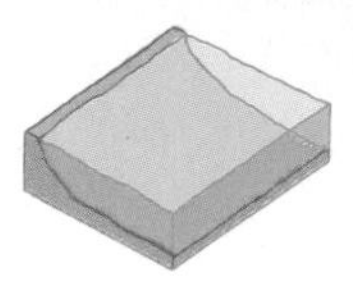

Classification Phylum Mollusca, Class Bivalvia, Order Veneroida, Family Veneridae.
Description The shell of *Venus* is oval, equivalved and inequilateral. It possesses an external ligament positioned behind the umbones; the hinge has three cardinal teeth on each valve and an anterior lateral tooth on the left valve. It is characterized by a flattened lunule and a smooth escutcheon. Inside, there are two muscle scars, almost equal in size, and the pallial line possesses a marked pallial sinus. The external ornament is provided by concentric lines accompanied by crests that make the surface of the shell very grooved. There may also be radial striae, but these are much less pronounced.
Stratigraphic position and geographical distribution The genus *Venus* first appears in the Oligocene (35 million years ago) and still occurs in seas today. It is found in fossil form in Europe, Africa, Asia and North America. The example in the photograph, which belongs to the species *V. multilamella*, comes from the Piedmontese Tertiary basin, more precisely from the Pliocene beds at Asti (Italy), and measures 3 cm (1¼ in).
Note *Venus* is a benthonic form that lives in sediment, from which only its two siphons emerge.

---

62 CONCHODON

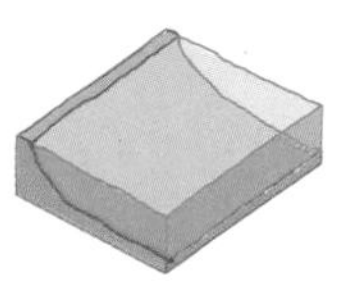

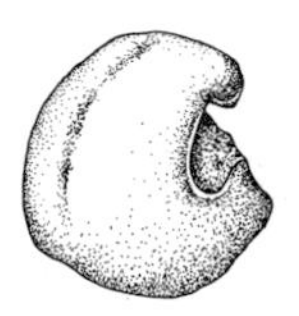

Classification Phylum Mollusca, Class Bivalvia, Order Hippuritoida, Family Megalodontidae.
Description The shell of the genus *Conchodon* is very swollen and triangular, almost oval, in shape. The umbones are very prominent and prosogyrous; the valves are characterized by a well-marked posterior keel. The hinge is composed, on the right valve, by a well-developed tooth, arched or irregularly tripartite, while on the left valve there is a pit accompanied by a horseshoe-shaped tooth; there are no secondary lateral teeth in the left valve. The interior reveals an adductor muscle scar, the posterior muscle scar is undetectable. The outer surface of the shell is characterized by thin and weakly defined concentric plicae.
Stratigraphic position and geographical distribution The genus *Conchodon* is found in fossil form solely in Rhaetian strata (Upper Triassic, 200 million years ago) and only on the continent of Europe. The example in the photograph, which belongs to the species *C. infraliasicus*, measures 10 cm (4 in) and comes from the Italian Rhaetian at Monte Prasanto.
Note There is a Rhaetian dolomite found in Italy which, because it contains abundant fossil remains of this bivalve, is called "*Conchodon* dolomite."

63 EDMONDIA

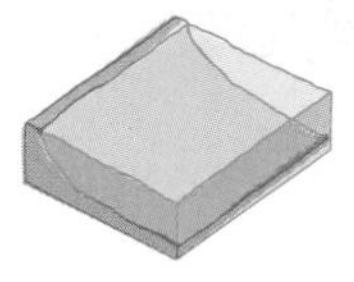

Classification Phylum Mollusca, Class Bivalvia, Order Pholadomyoida, Family Edmondiidae.
Description The shell of *Edmondia* varies in shape from oval to an elongate ellipse; it can also be characterized by a certain gibbosity. It has neither lateral ridges nor grooves; the umbones are very much reduced and anteriorly projecting. The shell is closed, equivalved and more or less inequilateral; the ligament is external and poorly developed. The interior reveals the scars of the muscles, which are almost equal in size, and a pallial line that is entire or marked by a negligible sinus. The hinge is extremely small.
Stratigraphic position and geographical distribution The genus *Edmondia* is found in strata ranging from the Upper Devonian (350 million years ago) to the Upper Permian (240 million years ago) throughout the world. The example in the photograph comes from the Carboniferous of Mazon Creek, Illinois, and measures approximately 2 cm (.75 in).
Note In the Mazon Creek beds the remains of this early bivalve are very frequent and are associated with plants and invertebrates.

---

64 GRYPHAEA

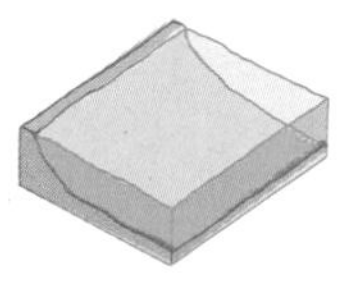

Classification Phylum Mollusca, Class Bivalvia, Suborder Ostreina, Family Gryphaeidae.
Description The shell of *Gryphaea* is medium-sized; the left valve is convex and there occurs a prominent, curving umbo on the right valve, which is much smaller than the left and has a shape ranging from concave to flat. The left valve is affixed to the substratum, while the right valve acts as an operculum. The ornament on the right valve is restricted to simple, thin growth lines, while the left valve is characterized by ornamentation composed of distinct growth lines accompanied by large numbers of lamellae. The hinge is very simple, with no dentition. The ligament area is characterized in its central section by the resilifer, which is sited under the umbo; there is also a very clear scar made by the sturdy adductor muscle.
Stratigraphic position and geographical distribution The genus *Gryphaea* appears in the Carnian (Upper Triassic, 210 million years ago) in Canadian, North American and Siberian sediments. Subsequently, during the Sinemurian (190 million years ago), it infiltrated all the seas of the world. It vanished in the Kimmeridgian (Upper Jurassic, 150 million years ago). The example in the photograph, which belongs to the species *G. navia*, comes from the Jurassic in Texas and measures 5 cm (2 in).
Note The abundant discoveries of this fossil in Pliensbachian limestone (Lower Jurassic) have meant that these limestones are now called "*Gryphaea* limestones." The habitat of this mollusc was the carbonate platform.

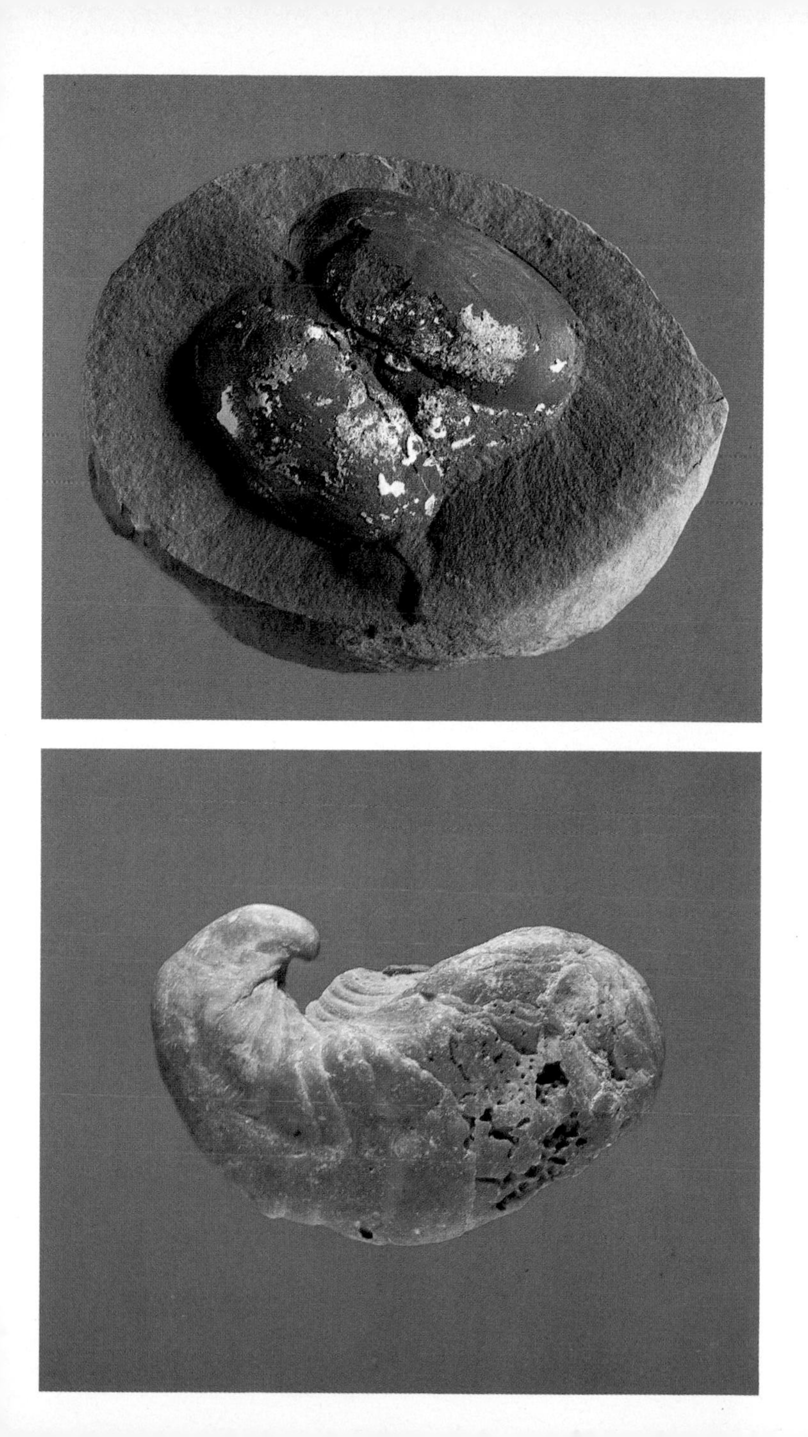

65 EXOGYRA

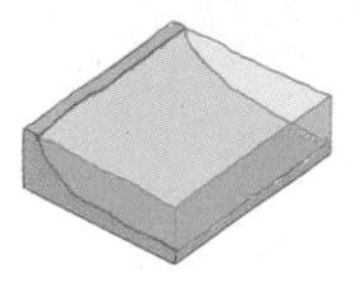

Classification Phylum Mollusca, Class Bivalvia, Suborder Ostreina, Family Gryphaeidae.
Description The shell of the genus *Exogyra* is medium-sized and strongly inequivalved: the right valve is concave or flat; the left valve is convex and characterized by an umbo that is much more pronounced than the one on the right valve, which rises up in a rolling spiral along its plane of commissure. Internally, there is an orbicular adductor muscle scar. Ornament is provided, on the right valve, by concentric, scaly striae running parallel to the anterior margin of the valve. The left valve is characterized by a broad plane of commissure bounded by a cord (chomata); its ornament consists of foliaceous growth scales, sometimes accompanied by radial costae that go on to form spines and tubercles.
Stratigraphic position and geographical distribution The genus *Exogyra* appears in the Lower Cretaceous (140 million years ago) and becomes extinct in the Upper Cretaceous (65 million years ago). Its area of fossil distribution embraces North America, Europe, northern and central Africa, Asia and India. The example in the photograph, which belongs to the species *E. costata*, comes from the Maastrichtian stratum in the state of Alabama and is approximately 65 million years old; it measures 7 cm (2¾ in).
Note The larva of this bivalve attaches itself to the substratum by the left valve; post-larval growth is spiral.

---

66 OSTREA

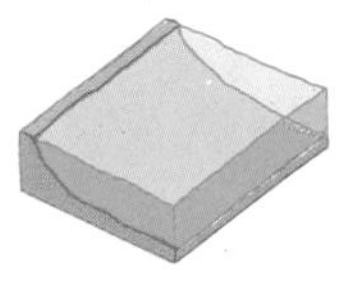

Classification Phylum Mollusca, Class Bivalvia, Suborder Ostreina, Family Ostreidae.
Description The shell assigned to the genus *Ostrea* is of medium size: thick and irregular in shape, it varies from the orbicular to the elongate. The left valve, which is attached to the substratum, is very thick and convex and therefore much more developed than the right one. The latter is fairly thin and may even appear flat. The resilifer is broad and takes the form of an elongate furrow, while there is no hinge (the hinge is present solely during the early stages of growth). Inside the shell there is an adductor muscle scar, very broad and kidney-shaped, positioned almost at the center of the valve. There is a visible pallial line, which is entire. The right valve is adorned with concentric growth lines, while the left one has an ornament of heterogeneous, radial costae accompanied by irregularly-arranged lamellae.
Stratigraphic position and geographical distribution The genus *Ostrea* appears in the Cretaceous (140 million years ago) and still occurs in seas today, except in polar regions. In the fossil state it enjoys a wide geographical distribution. The shell in the photograph comes from the Pliocene at Asti (Italy) and measures 12 cm (5 in).
Note The genus *Ostrea* can adapt itself to life both in turbulent, sandy beds (close to the coastline, for example, where the waves are most active) and on rocky or even muddy sea floors.

67 LECANOSPIRA

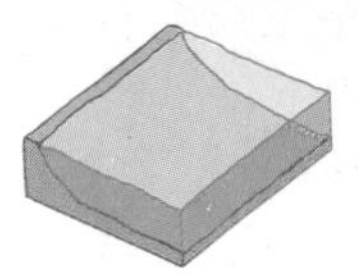

Classification Phylum Mollusca, Class Gastropoda, Order Archaeogastropoda, Family Macluritidae.
Description Gastropods included in the genus *Lecanospira* possess a modest-sized shell with planospiral coiling. Its general shape is discoidal, with a flat base and a broad, deep umbilicus. The peristome is olive-shaped: it is sharp in the upper region and ends in a strong crest, becoming rounded toward the basal region. The ornament of this gastropod consists of strong growth lines occurring solely on the upper region of the shell.
Stratigraphic position and geographical distribution The genus *Lecanospira* is a good guide fossil for Ordovician strata, where it is found in modest numbers almost throughout the world. The example in the photograph, 3 cm (1¼ in) in diameter, comes from the Ordovician (450 million years ago) of Spring Creek, Oregon.
Note The suborders Bellerophontina and Macluritina (to which the genus *Lecanospira* is assigned) and the suborder Pleurotomariina (to which the genus *Pleurotomaria* belongs) are the oldest groups of gastropods.

---

68 PLEUROTOMARIA

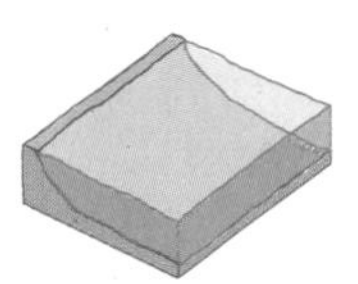

Classification Phylum Mollusca, Class Gastropoda, Order Archaeogastropoda, Family Pleurotomariidae.
Description The shell of *Pleurotomaria* is trochiform, with a moderately-raised, clockwise-coiling spire. The right margin of the aperture is characterized, in its central section, by a deep, straight fissure that is gradually closed together by the growth of the shell, a process giving rise to a peculiar scar, sometimes known as a "fasciole," along the successive whorls. This cleft allows for the expulsion of water circulating inside the mantle. The whorls occur in small steps, flat or gently concave. The ornament of this fossil is characterized by thick, spirally-arranged bands accompanied by tubercles; this tubercular ornament may also occur on the lower section of the whorl.
Stratigraphic position and geographical distribution The genus *Pleurotomaria* appears in Lower Jurassic strata (190 million years ago) and becomes extinct in the Lower Cretaceous (120 million years ago). Its distribution is worldwide. The example in the photograph, 2 cm (¾ in) in diameter, was discovered in Austria in Cretaceous sediments.
Note The genus *Pleurotomaria* was widespread during the Middle Mesozoic and was characteristic of warm, oxygenated marine waters. The diet of this gastropod was herbivorous.

69 FISSURELLA

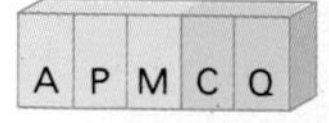

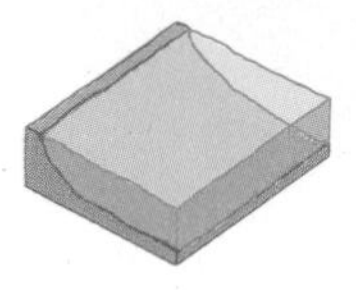

Classification Phylum Mollusca, Class Gastropoda, Order Archaeogastropoda, Family Fissurellidae.
Description The general shape of this curious gastropod is conical and reminiscent of that of the genus *Patella*. The shell is very thick and possesses a spiral protoconch. It is characterized by the presence, near the apex, of a fissure accompanied by a rough, rounded collar; this fissure serves to expel water from the interior of the shell. Below is the peristome, which occupies the entire lower part of the animal. The external surface of *Fissurella* displays tight, thick costae, radially arranged, and concentric growth rings. There are also thin striae visible on the interior of the shell. The shell's interior also reveals a characteristic horseshoe-shaped imprint made by the foot, which takes the form of two parallel branches sited in the anterior region of the animal.
Stratigraphic position and geographical distribution The genus *Fissurella* is a poor guide fossil since it is known since the Eocene (55 million years ago) in sediments throughout the world and still exists in very similar forms today. The example in the photograph, 3 cm (1¼ in) in diameter, comes from the Pliocene at Asti (Italy).
Note Gastropods of the genus *Fissurella* now live in warm, tropical seas with a very high degree of oxygenation. Discovery of this fossil allows us to reconstruct with relative accuracy the paleoclimatic conditions in which its surrounding sediments were laid down.

---

70 PATELLA

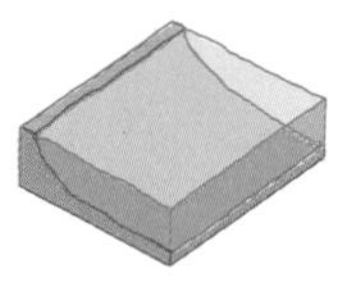

Classification Phyllum Mollusca, Class Gastropoda, Order Archaeogastropoda, Family Patellidae.
Description The shell of *Patella* is conical, generally very thick and with bilateral symmetry. Its base may be roundish or, in some species, oval leading to octagonal; the apex can be central or slightly displaced. The protoconch is not visible. The general ornament is provided by prominent ribs radiating from the apex; the arrangement of these ribs is completely irregular and sometimes they are accompanied by thin, intersecting growth lines. Internally, the shell reveals a pronounced muscle scar (left by the foot) in a central position. Unlike other gastropods, the shell shows no tendency towards spiral development, a feature that indicates the archaic character of the species of this genus. On the other hand, the larval stage of this gastropod shows clear spiralism.
Stratigraphic position and geographical distribution The genus *Patella* definitely occurs in sediments throughout the world since the Eocene (55 million years ago) and is still found in seas today. The attribution to this genus of fossil finds made in Cretaceous sediments is open to doubt. The example in the photograph, which belongs to the species *P. italica*, comes from Pliocene sediments at Asti (Italy); it measures 3 cm (1¼ in).
Note *Patella* attaches itself by means of its powerful foot to either rocky substrata or other, rigid-bodied animals. It is a herbivorous gastropod that feeds on algae.

71 TROCHUS

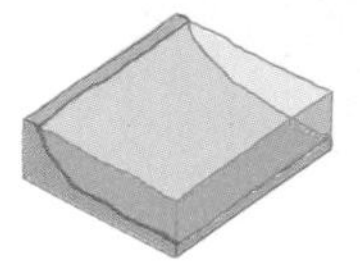

Classification Phylum Mollusca, Class Gastropoda, Order Archaeogastropoda, Family Trochidae.
Description The shell of the genus *Trochus* is conical and composed of raised, flat whorls. The final whorl is angular at the periphery of the base, which is flattened and concave at the center. The aperture is rhomboidal, with a slender, discontinuous peristome whose opposite edges do not lie on the same plane; the outer lip is thin. The columella twists backwards and is established on a false umbilicus covered by a callus. The ornament can be siphonal or collabral. There are sometimes numerous tubercles or granules.
Stratigraphic position and geographical distribution The earliest known examples of the genus *Trochus* are derived from fossil-bearing deposits of the Miocene (20 million years ago) and it still occurs in numerous forms in seas today. The example in the photograph, 2.5 cm (1 in) high, comes from Pliocene strata at Bordeaux (France).
Note Like other representatives of the family, the genus *Trochus* is equipped with a round and horny operculum.

---

72 TURBO

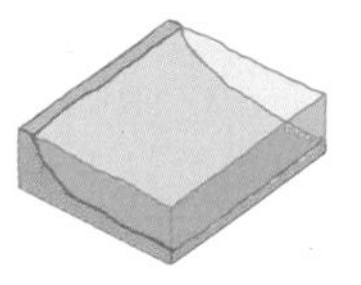

Classification Phylum Mollusca, Class Gastropoda, Order Archaeogastropoda, Family Turbinidae.
Description The genus *Turbo* possesses shells with a typically turbinate shape and the animals may sometimes reach considerable sizes. The final whorl is very dilate and convex; the whorls are generally not very convex and are adorned with wrinkles that may also possess nodules and tubercles, or the ornament may be characterized by spiral threads separated by shallow striae. The aperture is very broad and has a circular shape overall; the genus *Turbo* is also distinguished by a moderately developed operculum with a flat internal surface. There is a highly developed columella.
Stratigraphic position and geographical distribution The genus *Turbo* occurs in its earliest fossil form in Upper Cretaceous sediments (80 million years ago) and is found widely in Europe, Asia, North America and South America, as well as in Australia and Africa. It is still widespread today. The example in the photograph, *T. rugosus*, comes from the Quaternary at Dahlia, Zimbabwe; it measures 2.5 cm (1 in) in diameter.
Note The genus *Turbo* is now characteristic of warm seas with oxygenated waters and is therefore a good paleoecological indicator.

73 TURRITELLA

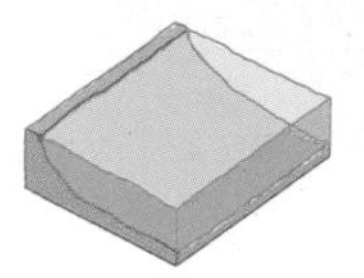

Classification Phylum Mollusca, Class Gastropoda, Order Mesogastropoda, Family Turritellidae.
Description The shell of this gastropod is conical, typically turreted and thin. The spire is very elongate and its whorls are convex and adorned with large numbers of spirally arranged striae accompanied by very thin growth lines curving toward the lower suture. The sutures are deep and linear, always slightly slanting and surmounted by a faint keel. The last whorl differs very little from the others and sometimes displays a slightly angular ending. The aperture is oval or anteriorly subangular; the lip is thin and the columella smooth and curved. This gastropod possesses a horny operculum.
Stratigraphic position and geographical distribution The genus *Turritella* is first known from strata of the Upper Cretaceous (80 million years ago) and still occurs in seas today. It is found in fossil form all over the world. The example in the photograph, which belongs to the species *T. turris*, is 4 cm (1½ in) high and comes from the Italian Pliocene at Asti.
Note The greatest evolutionary development of this genus took place during the Eocene, but it makes an excellent guide fossil for the whole of the Tertiary. It is a herbivorous form that favors soft sea floors.

---

74 NATICA

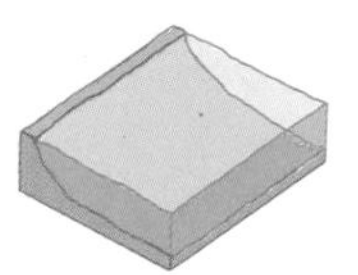

Classification Phylum Mollusca, Class Gastropoda, Order Mesogastropoda, Family Naticidae.
Description The shell is fairly large in size; it is of globular form, with a very short spire. The whorls are fairly few in number and grow rapidly. The sutures are linear, sometimes equipped with a spiral keel. The surface of this gastropod is smooth and displays small brown speckles. The last whorl is rounded and forms the largest part of the shell: it is very broad and reveals a similarly broad umbilicus. Within the umbilicus the funicle is sometimes visible. The aperture is half-moon-shaped and is placed at a level of 40 or 50 degrees in relation to the general axis of the shell. The axial part of the aperture, which is rectilinear, bears a callus of varying dimensions that may completely obscure the umbilicus. The ornament of this gastropod is provided by light striae (which are not always present). The aperture possesses a calcareous operculum, half-moon-shaped like the aperture, which is sometimes used systematically to identify individual species.
Stratigraphic position and geographical distribution The genus dates from the Cretaceous and is still found today. Its area of distribution is vast. The example in the photograph, which measures 2.5 cm (1 in) in diameter, comes from the Florida Pliocene (5 million years ago).
Note Examples of *Natica* discovered in the European Miocene still reveal traces of color.

75 MELANOPSIS

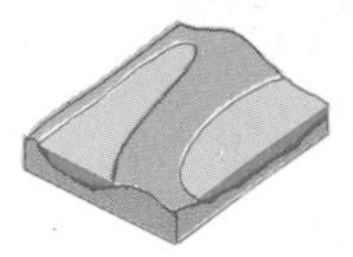

Classification Phylum Mollusca, Class Gastropoda, Order Mesogastropoda, Family Melaniidae.
Description The shell of *Melanopsis* is of modest size and either conico-ovoid or conico-fusiform. The spire is moderately elongate and there are only a few whorls, convex or flattened, which overlap to a third of their height. The last whorl is of the same size or equal to two-thirds of the total height of the shell. The aperture is small, oval and partly covered posteriorly by the last whorl. The columella is smooth and curving, with a very rough edge at the posterior end. The ornament is provided by thin growth lines and may possess prominent or faint longitudinal ribbing; in exceptional cases the surface of the shell may be smooth. When ribs are present, these may be accompanied by nodules and spines.
Stratigraphic position and geographical distribution The genus *Melanopsis* appears in the Upper Cretaceous (80 million years ago) and is still found in seas today. It is characteristic of Tertiary sediments both in Europe and elsewhere. The example in the photograph, 1.5 cm (½–¾ in) high, comes from the French Eocene at Cuise-la-Motte.
Note This gastropod is a good paleoecological indicator since its modern forms live in fresh or brackish waters.

---

76 STROMBUS

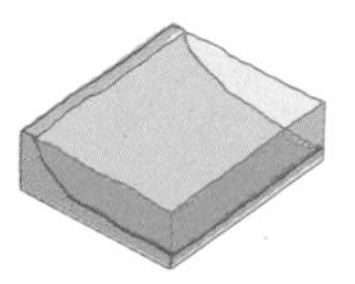

Classification Phylum Mollusca, Class Gastropoda, Order Mesogastropoda, Family Strombididae.
Description The shell of *Strombus* is very strong. This is a large gastropod and it is not uncommon to find giant-sized specimens. The general form of the shell is biconical, with one of the two cones much more developed than the other. The spire is generally conical; some forms may display a concave outline. The shell is composed of few whorls, with the last one, particularly large, representing between two-thirds and three-quarters of the total height. This last whorl is expanded, with an oval base slightly hollow in the posterior section, where a large callus occurs. The aperture is elongate, with parallel edges and a large, well-developed inlet. The stout columellar lip, extended and widely dilated, is marked by thin striae.
Stratigraphic position and geographical distribution The genus *Strombus* has been known since the Eocene (55 million years ago) and still occurs in seas today. It has a broad geographical distribution. The example in the photograph, 10 cm (4 in) high, comes from the Italian Pliocene at Asti.
Note The species *S. bubonius* is a good guide fossil for the Tyrrhenian (Upper Pleistocene, 100,000 years ago).

77 MELONGENA

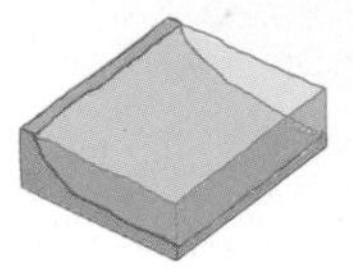

Classification Phylum Mollusca, Class Gastropoda, Order Neogastropoda, Family Fusidae.
Description *Melongena* is characterized by a large, very strong shell. The general shape of the shell ranges from oval to biconical and appears swollen. The spire is short, with a conical outline, and the whorls are narrow and angular toward the front. In this angular region numerous ridged nodules are found, and these may become transformed into very sharp and well-developed spines on the surface of the last whorl, where their conjunction creates a low, spiny and continuous corona. The aperture is pyriform, very high (inasmuch as the last whorl is elongate and extremely dilated), broad and spacious, with an ornate columellar lip. The lip is narrow, generally smooth or faintly grooved, while the columella is posteriorly shallow and sometimes slightly convex.
Stratigraphic position and geographical distribution *Melongena* is a typical genus from the Tertiary. It ranges from the Middle Eocene (a little less than 55 million years ago) to the Miocene, when it became extinct. The example in the photograph, 7 cm (2¾ in) high, comes from the Italian Eocene at Altavilla, Vicenza.
Note This was a littoral, benthonic animal, which lived in subtropical waters. It was a carnivore enjoying a high degree of mobility thanks to its large foot. *Melongena* is a good guide fossil.

---

78 OLIVA

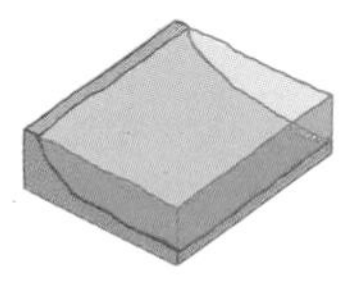

Classification Phylum Mollusca, Class Gastropoda, Order Mesogastropoda, Family Olividae.
Description The genus *Oliva* has a medium-sized shell bearing a relatively low spire and a last whorl that almost completely envelops the preceding one. This curious gastropod takes its name from the general shape of its shell, which recalls that of an olive. The suture is incised in all the whorls, except the last one, in which it is cuniculate. The aperture is narrow and elongate, with a narrow, posteriorly-sited inlet into which the pallial siphon is inserted. The outer lip is smooth, while the columellar lip bears a callus characterized by the presence of thin saddles accompanied by denticulation or grooves. Members of *Oliva* are distinguished by their lack of surface ornament.
Stratigraphic position and geographical distribution Representatives of the genus *Oliva* are found throughout the world in sediments of the Upper Cretaceous (80 million years ago), and it still enjoys a wide geographical distribution. The example of *O. savana* in the photograph, .5 cm (¼ in) high, comes from the Florida Pliocene.
Note *Oliva* is regarded as a good paleoecological indicator because it favors warm, shallow waters with good oxygenation. It lives both in and out of the sand.

79 FICUS

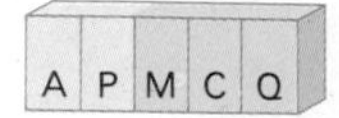

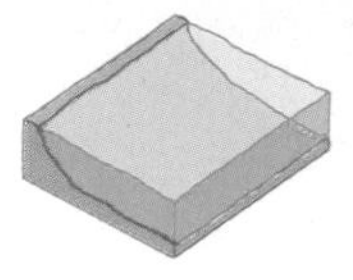

Classification Phylum Mollusca, Class Gastropoda, Order Mesogastropoda, Family Ficidae.
Description The shell of the genus *Ficus* is very swollen, with a shape reminiscent of a fig. It coils in an open spiral with a narrow spire whose tip is barely distinguishable from the apex of the shell. The last whorl is highly dilated, particularly in the upper part, while in the basal section it narrows to form a long siphonal canal that diverges slightly from the shell's general axis. A more or less axially developed siphonal canal is formed. The ornament is provided by numerous spinal bands, which are in turn accompanied by less prominent spiral structures that intersect other, radial structures, thereby giving the shell's surface a reticular appearance.
Stratigraphic position and geographical distribution The genus *Ficus* first appears in the Upper Cretaceous (80 million years ago) and still occurs in seas today. The example in the photograph, which belongs to the species *F. papyratia*, comes from the Florida Pliocene and is 2 cm (.75 in) high.
Note The genus occurs widely in sediments of the Miocene and Pliocene, for which the species *F. ficoides* is a good guide fossil.

---

80 MUREX

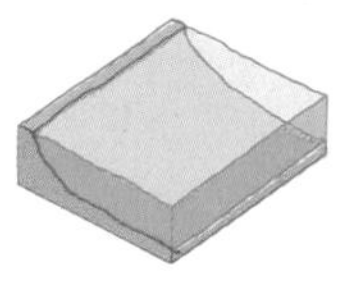

Classification Phylum Mollusca, Class Gastropoda, Order Neogastropoda, Family Muricidae.
Description The shell ranges in shape from oval to oblong and its surface is "muricate," meaning that it bears growth lamellae that create rough projections in varying densities. It is solid in form, with the overall shape of a club. The spires are relatively short and the protoconch is smooth. The whorls are angular, the angles bearing six tubular spines that correspond to the axial varices. The last whorl is large and generally adorned with spiral striae. Its base is slightly convex and gouged out by the growth of the collar, which is very elongate and equipped with a spiny pad at the intersection of the extremities of the axial varices. The aperture is oval and the lip is usually vertical, thickened by a callus; the columella is regular and posteriorly smooth. The edge of the columella is visible.
Stratigraphic position and geographical distribution The genus *Murex* first appears in the Miocene (20 million years ago) and is still present in seas today. In the past it occurred, as it does today, in warm waters throughout the world. The example in the photograph, 4 cm (1½ in) high, comes from the Italian Pliocene at Altavilla, Vicenza.
Note *Murex* and the Muricidae generally favor firm, rocky sea bottoms or reefs. Gastropods of the genus *Murex* are good paleoecological indicators.

81 NASSA

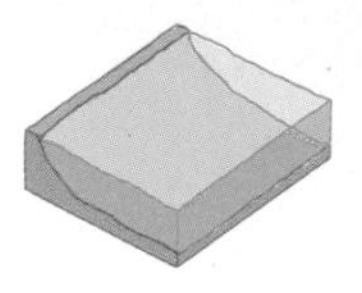

Classification Phylum Mollusca, Class Gastropoda, Order Neogastropoda, Family Nassariidae.
Description Shells of the genus *Nassa* are fairly small and thin-walled, with an oval, very swollen shape. The spire is moderately elongate, with a conical outline; the protoconch is smooth and subglobular. The whorls are convex, running at right angles to the suture, and are often smooth or slightly ribbed. The last whorl is large, oval and rounded at the base, which invariably has a groove made by the siphon. The aperture is oval with a narrow, abapically retroflex sinus. Folds occur inside the outer lip of the aperture. Ornament is provided by elements running either spirally or collabrally, although occasionally there are smooth forms.
Stratigraphic position and geographical distribution The genus *Nassa* dates from the Eocene (55 million years ago) and still occurs widely in the seas today. The example in the photograph belongs to the species *N. conglobata* and comes from the Italian Pliocene at Asti; it has a diameter of 2.5 cm (1 in).
Note During the Pliocene this genus became very widespread and many different species of it existed.

---

82 CONUS

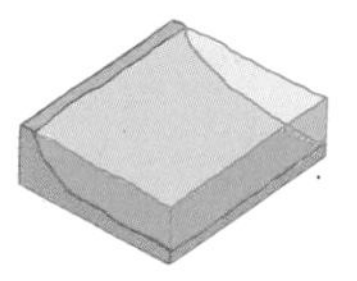

Classification Phylum Mollusca, Class Gastropoda, Order Neogastropoda, Family Conidae.
Description The genus *Conus* has a biconical shell with a surface that is either smooth or decorated with spiral striae. The spiral has tightly packed whorls, always shorter than the aperture, which is straight, has parallel edges and ends anteriorly in a siphonal canal. The lip is thin, not arched and is marked near the suture by a kink coinciding with the lower edge of the last whorl. The border of the columella is rectilinear and displays a fairly prominent fold near the front and also a small parietal groove in the lower corner.
Stratigraphic position and geographical distribution The genus *Conus* is known from the Upper Cretaceous (80 million years ago) and was discovered in Californian sediments. It is now found in seas in the Indo-Pacific region. The example in the photograph, *C. antediluvianus*, is 5.5 cm (2¼ in) high and comes from Pliocene sediments at Modena (Italy).
Note This is a carnivorous gastropod that lives on sandy sea bottoms; there are numerous known fossil species, some of which reach considerable sizes. Some species of the genus *Conus* are used as guide fossils during the Pliocene.

83 ACTAEONELLA

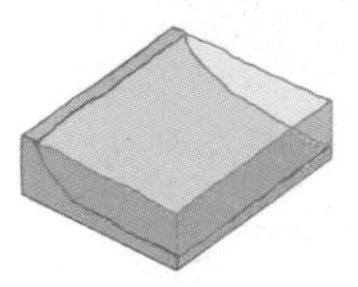

Classification Phylum Mollusca, Class Gastropoda, Order Pleurocoela, Family Orthostonidae.

Description The genus *Actaeonella* has a subcylindrical or ovoconical shell with totally involute spires. The thickness of the shell is particularly marked. The last whorl is very developed and there is no trace of ornament on the external surface. The aperture defined by the last whorl possesses parallel edges, even though there is a noticeable dilation centrally. The lip is thin and almost straight for its entire length. The gently curving columella is decidedly short and adorned by three almost horizontal folds, which create grooves running from front to back.

Stratigraphic position and geographical distribution The genus *Actaeonella* is found in abundance in Upper Cretaceous sediments (80 million years ago) throughout most of the world. The example in the photograph, which is 4.5 cm (1¾ in) in diameter, comes from the German Cretaceous at Brandenburg.

Note Shells of *Actaeonella* are found in association with *Hippurites* and *Nerinea*, suggesting that this gastropod inhabited a reef environment with warm, oxygenated waters.

---

84 PLANORBIS

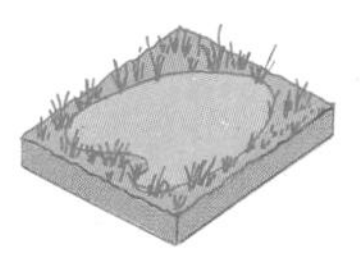

Classification Phylum Mollusca, Class Gastropoda, Order Basomatophora, Family Planorbidae.

Description The shell of *Planorbis* is of small dimensions and planospiral in form, although it sometimes tends slightly toward the trochoid; the coiling is sinistral. The spire is loosely coiled and the last whorl may be either dilated or follow the same general course as that of the other whorls. Seen from above, the shell of *Planorbis* is flat or slightly concave (like a flat-bottomed basin), while its lower part may be either flat or gently curving. The peristome is circular and never has horny collars. The ornament of this shell is provided only by weak growth striae, most of the time so faint that the surface appears practically smooth. Only rarely are the growth lines accompanied by insignificant spiral striae.

Stratigraphic position and geographical distribution The genus *Planorbis* is found in fossil form in Tertiary sediments (starting 50 million years ago) and still exists today. The earliest genera belonging to this family have been discovered in the Jurassic of France. The photograph shows an example of *P. complanata*, 1.5 cm (½–¾ in) in diameter, from the Quaternary of Ulm (Germany).

Note The genus *Planorbis* is a freshwater, pulmonate mollusc. It is a good paleoecological indicator.

85 ATURIA

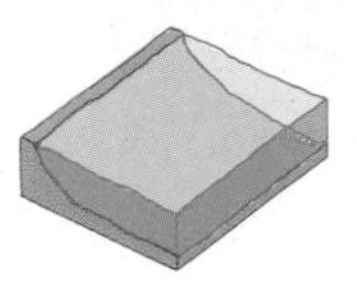

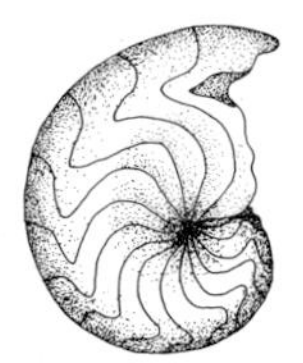

Classification Phylum Mollusca, Class Cephalopoda, Subclass Nautiloidea, Family Aturiidae.
Description An extinct nautiloid with a planospiral, highly involute shell and very small umbilicus. The sides are flattened, while the ventral region (the outermost one) is rounded. The shell is smooth in appearance; on fossil forms, however, the surface is enlivened by the geometrical pattern of the suture lines. Characteristically, the suture is composed of complete and very pronounced saddles and lobes, the latter sharp and the former broad and rounded. The siphuncle is dorsally sited (in the innermost part of the coil).
Stratigraphic position and geographical distribution *Aturia* is a typical Tertiary nautiloid that lived from the Paleocene to the Miocene (65–7 million years ago) and occurs widely throughout the world. The example in the photograph, which measures 6 cm (2½ in) in diameter, comes from the Italian Oligocene at Ovada in Piedmont.
Note Some examples of *Aturia* have retained their original color: this consists of traces of brown, similar to those found in the modern *Nautilus*, whose coloring is made up of splashes of brown on a white ground. *Nautilus*, the sole modern cephalopod to possess an external shell, is one of the best known of the "living fossils," a term used to describe those primitive organisms that have survived until the present.

---

86 DOLORTHOCERAS

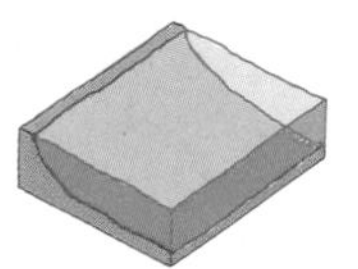

Classification Phylum Mollusca, Class Cephalopoda, Subclass Nautiloidea, Order Orthocerida, Family Pseudorthoceratidae.
Description A typical form of orthocone Paleozoic nautiloid: the shell is elongated, circular in section or slightly compressed, with an overall shape resembling that of a long cone flaring gently in the direction of growth. It is transversally crossed by sutures, which can be either straight or very slightly slanting. The siphuncle is centrally situated during the early stages of development, but in many species it later comes to lie toward the ventral region. The shell bears no ornament.
Stratigraphic position and geographical distribution The genus ranges from the Upper Ordovician to the Lower Permian (400–260 million years ago). It is found in North America, Europe, Asia and Australia. The example in the photograph, which belongs to the species *D. sociale*, comes from the Ordovician of Maquoketa Creek, in the state of Iowa.
Note Nautiloidea are seldom found in abundance: a rare exception is the large accumulations of shells of *D. sociale* that occur in Ordovician rocks in Iowa, where they form a real coquina or shell bed. The order Orthocerida, whose members were essentially Paleozoic (though they are found up until the Triassic), contains examples that can reach 2 meters (6½ ft) in length, although this is less than that achieved by other "uncoiled" nautiloids, such as Actinoceridae (6 meters or 19½ ft) and Endoceridae (9 meters or 29½ ft).

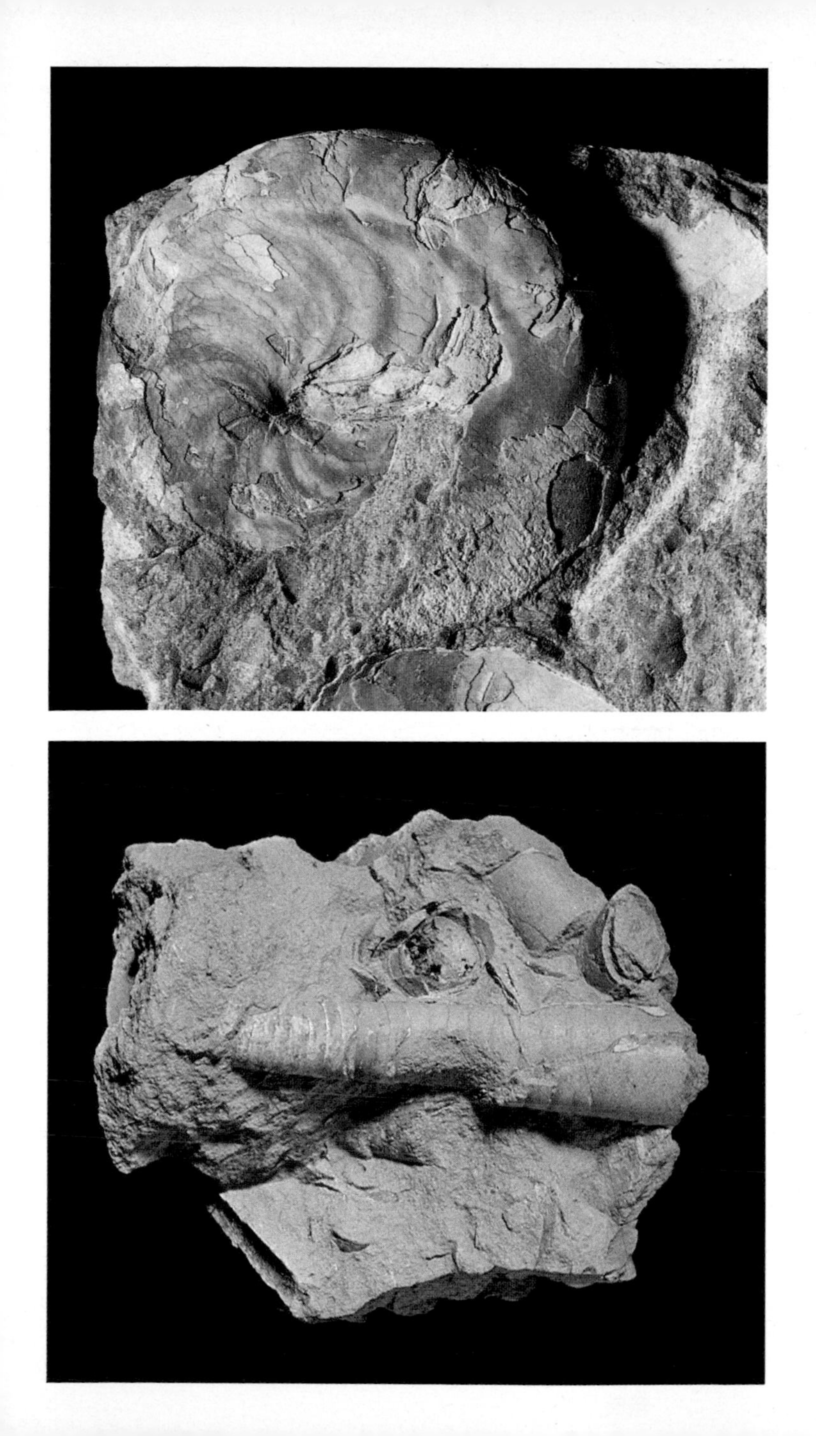

87 LITUITES

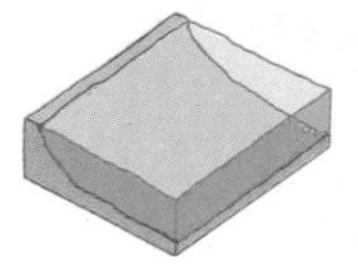

Classification Phylum Mollusca, Class Cephalopoda, Subclass Nautiloidea, Order Tarphycerida, Family Lituitidae.
Description A nautiloid with a shell only partially coiled in a flat spiral (lituiticone): the coiling appears in the first whorls, which may be only slightly touching, and the shell subsequently develops in a straight line, only very slightly sigmoid in form. The animal's body chamber occupies the whole of this uncoiled part of the shell. The shell aperture of adult specimens possesses two pairs of lateral expansions, one placed ventrally and the other dorsally. The siphuncle is situated in a dorsal position. The ornament of the shell is composed of thin, transversal growth lines, which are superimposed on ribs formed by broad transversal swellings that follow a more or less winding course.
Stratigraphic position and geographical distribution The genus is known from the Llanvirn to the Caradoc stages (Middle Ordovician, 475–440 million years ago). It is typical of the Baltic and Scandinavian countries. The example in the photograph belongs to the type species *L. lituus* and measures approximately 12 cm (5 in); it comes from the Swedish Ordovician.
Note Representatives of Tarphycerida, which are fairly rare in fossil form, range from the Lower Ordovician to the Upper Silurian; they include the earliest forms of cephalopod with shells beginning in a coil. The family Lituitidae has a restricted distribution, occurring only in northern Europe.

---

88 BACTRITES

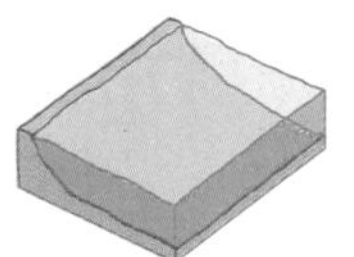

Classification Phylum Mollusca, Class Cephalopoda, Subclass Bactritoidea, Order Bactritida, Family Bactritidae.
Description A cephalopod with a typical orthocone shell, narrow and elongate in form, which broadens in the direction of growth; it is round or oval in section. The sutures are straight, transversal to the shell, with a small ventral lobe. The shell is ornamented with transversal growth lines that form a sharp, forward-pointing convexity in the ventral area.
Stratigraphic position and geographical distribution The genus is known with certainty from the Lower Devonian to the Upper Permian (395–240 million years ago). Its geographical distribution is wide: central Europe, the Urals, Sicily, Morocco, Australia, the U.S., Mexico and Peru. The example in the photograph, approximately 6 cm (2½ in) long, comes from the German Upper Devonian at Adorf in the Eifel and belongs to the species *B. gracilis.*
Note Bactritidae are cephalopods whose systematic position has long been the subject of debate because they show characteristics intermediate between Nautiloidea and Ammonoidea. They were a Paleozoic group, first known in the Ordovician and becoming totally extinct at the end of the Permian, the period marking the end of the Paleozoic era.

89 GONIATITES

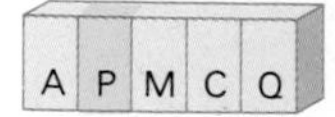

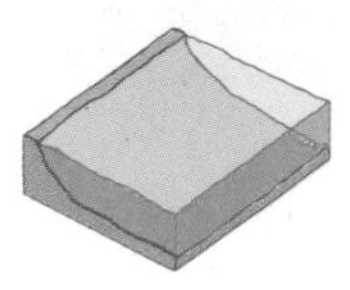

Classification Phylum Mollusca, Class Cephalopoda, Subclass Ammonoidea, Family Goniatitidae.
Description Spirally coiling shell, very involute, with the last whorl covering almost all the preceding whorls; the umbilicus is generally very narrow as a consequence, although in certain forms it is somewhat broader. The ornament consists of thin growth lines. The shell's aperture has a characteristic, forward-turning embayment on its ventral margin (the outermost section). The suture line is of the very simplified "goniatitic" form: it displays entire lobes and saddles, with an acute, elongate lateral lobe and a ventral lobe interrupted by a saddle. Seen in longitudinal section, the shell reveals a ventrally positioned siphuncle, as in Mesozoic Ammonoidea, but still bears septal necks (which encircle the siphuncle where it passes through the septa) that project backwards instead of forward, as in the more evolved Ammonoidea.
Stratigraphic position and geographical distribution The genus is typical of the Visean stage of the Carboniferous (345 million years ago) and is found in North Africa, Belgium, Germany and England. The example in the photograph comes from the Moroccan Carboniferous at Erfoud; it measures 4.5 cm (1.75 in).
Note The fossil remains of *Goniatites* are typical of sediments formed in shallow coastal waters; the genus required water that was warm and well oxygenated by the presence of strong currents.

---

90 BISATOCERAS

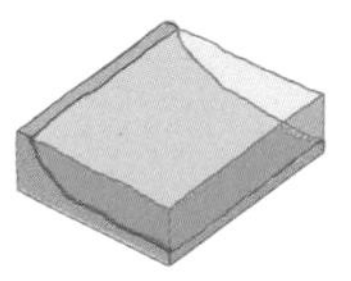

Classification Phylum Mollusca, Class Cephalopoda, Subclass Ammonoidea, Family Goniatitidae.
Description The shell is of almost globular form, with a subcircular section and a very rounded ventral area. The last whorl mostly covers the preceding ones and the umbilicus is consequently narrow and very deep, with a well-accentuated edge and straight or reentrant walls. The ornamentation is faint and composed of thin, regular transversal ribs. The suture is goniatitic, very simple in form, with a broad and slightly accentuated umbilical saddle, a distinct lateral lobe and a bifid, equally pronounced ventral lobe.
Stratigraphic position and geographical distribution The genus is typical of the Lower and Upper Pennsylvanian strata in the central and southwestern U.S., a period that corresponds to the European Upper Carboniferous (300 million years ago). The example in the photograph, which belongs to the species *B. greenei*, comes from the Pennsylvanian in Okmulgee County, Oklahoma, and measures approximately 3 cm (1¼ in).
Note *Bisatoceras* belongs to the subfamily Bisatoceratinae, which also includes the genera *Peenoceras*, which occurs in the Pennsylvanian in Ohio and Pennsylvania, and *Nuculoceras*, which occurs in the English Lower Carboniferous. All are characterized by a very narrow umbilicus and a bifid ventral lobe on the suture line.

91 CERATITES

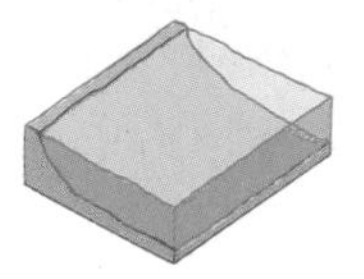

Classification Phylum Mollusca, Class Cephalopoda, Subclass Ammonoidea, Family Ceratitidae.
Description A somewhat evolute, planospiral shell, with a large umbilical opening—laterally compressed, the height of the whorls is greater than their width—and decorated with large radial costae sometimes forming nodes around the umbilical margin. The body chamber—the terminal part of the shell within which the animal lived—is short. The simple "ceratitic" suture line has broad, rounded entire saddles and no denticulation. The lobes, however, are finely toothed and narrower than the saddles. The siphuncle is ventral with prosiphonate septal necks.
Stratigraphic position and geographical distribution The genus is characteristic of the European Lower Triassic, particularly the so-called "Muschelkalk." The type species is *C. nodosus*, an example of which, from the German Muschelkalk, is shown in the photograph, measuring 6 cm (2½ in).
Note *Ceratites* abounds in the shallow marine sediments of the central European Muschelkalk. These sediments were laid down in brackish waters inhabited by a few animal species, each with numerous members. Remains typically associated with those of *Ceratites* are fossilized Brachiopoda and Crinoidea of the species *E. lilliformis*. Members of the suborder Ceratitina, to which *Ceratites* belongs, were very widespread in the Triassic, a period for which they are of considerable stratigraphic importance. No representative of the order, however, survived from the Triassic to the Jurassic.

---

92 CLADISCITES

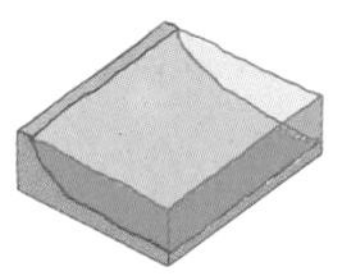

Classification Phylum Mollusca, Class Cephalopoda, Subclass Ammonoidea, Family Cladiscitidae.
Description An ammonite of a relatively unusual shape: its shell, which has a flat coil, possesses a coil section that is rectangular, with a greater width in the outermost or "ventral" region. This region is very broad and, like the sides, rather flattened. The whorls overlap greatly, resulting in a very narrow umbilicus. The ornament consists of striae running parallel to the direction in which the shell is coiled, and there are no nodes or other types of ribbing. The suture, although this is a Triassic genus, is of the ammonitic type, with an intricate structure involving bifid saddles and very frilled lobes.
Stratigraphic position and geographical distribution The genus is widespread from the Carnian to the Rhaetian (Upper Triassic, 210–190 million years ago) and has a vast geographical distribution, occurring in the Alps, in Sicily, the Balkans, the Himalayas, Timor and in the state of Alaska. The example in the photograph, which belongs to the species *C. crassestriatus*, comes from the island of Timor and measures 7 cm (2¾ in).
Note Because of its broad geographical distribution, the genus is used as a guide fossil for the Upper Triassic, particularly the species *C. ruber*, which occurs only in the Norian.

93 PHYLLOCERAS

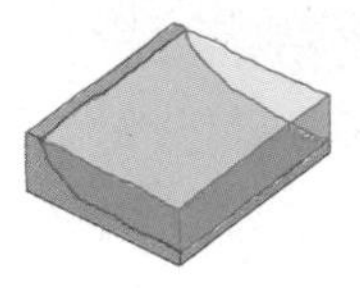

Classification Phylum Mollusca, Class Cephalopoda, Subclass Ammonoidea, Family Phylloceratidae.
Description An involute, planospiral shell, with a very narrow umbilicus; the coil section is oval and increases very rapidly in height though much less so in width, creating its typically flattened discoidal shape, with a rounded ventral area. Ornament is either absent or very faint: where it exists, it consists of simple growth striae that break in the ventral region at the point where the siphuncle occurs. The suture is very complex: it consists of lobes and narrow, elongate saddles extensively indented and subdivided into secondary lobes and saddles, thus forming a surface network of marks that look like a dense tangle of leaves. The body chamber occupies less than half of the last whorl.
Stratigraphic position and geographical distribution *Phylloceras* extends over a long period of time, which is unusual in Mesozoic ammonites: it first appears at the beginning of the Jurassic and through numerous very similar species, survives until the Cretaceous (190–130 million years ago). It is often found in many European layers. The photograph shows *P. doederlanium* from the Italian Toarcian (Lower Jurassic) at Furlo in the Apennines; it measures 8 cm (3¼ in).
Note The Phylloceratidae represent one of the three great groups of ammonites characteristic of the Jurassic and Cretaceous. First appearing in the Triassic, they were the only ammonites to survive from the Triassic to the Jurassic.

---

94 LYTOCERAS

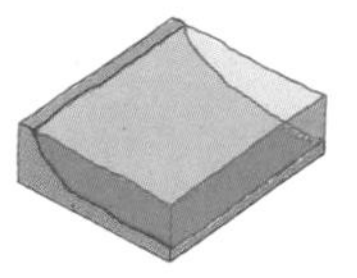

Classification Phylum Mollusca, Class Cephalopoda, Subclass Ammonoidea, Family Lytoceratidae.
Description Very evolute shell, with whorls ranging in shape from more or less rounded to almost square, and either slightly overlapping or just touching. The umbilicus is consequently very broad and shallow, with rounded umbilical walls and a rounded, unclear umbilical margin. The ornament is generally provided by simple, radial-growth striae, which are straight, thin and regular. The suture is ammonitic, with very heavily frilled lobes and saddles. There are two bifid lateral lobes; the outermost one is more pronounced. The ventral lobe, also very pronounced, is similarly bifid.
Stratigraphic position and geographical distribution The genus *Lytoceras* survives for a considerable length of time, stretching from the Sinemurian (Lower Jurassic) to the Upper Cretaceous (190–70 million years ago). It enjoys a wide geographical distribution, occurring throughout the world, even at high latitudes, with examples of *Lytoceras* being found as far nor as Greenland and northern Alaska. The example in the photograph, of the species *L. fimbriatum*, comes from the Italian Domerian (a stage of the Pliensbachian, Lower Jurassic, 180 million years ago) at Mount Domaro in the Alpine foothills of Lombardy; it measures 6 cm (2½ in).
Note Because of the shape of its shell, it is believed that *Lytoceras* was a nektonic form, capable of swimming. It is abundant in layers in considerable numbers and specimens can be large, with shells measuring dozens of centimeters.

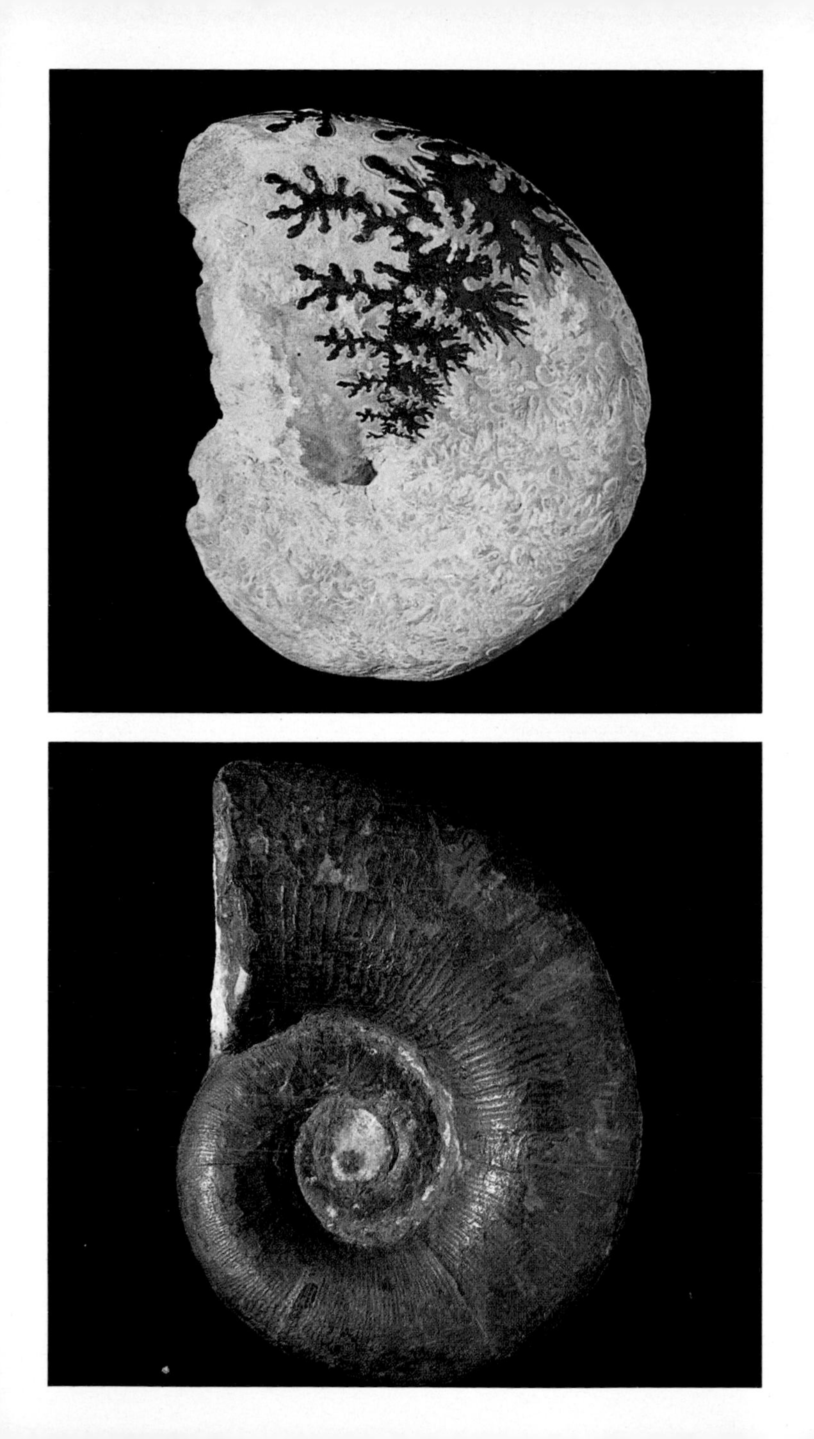

95 MACROSCAPHITES

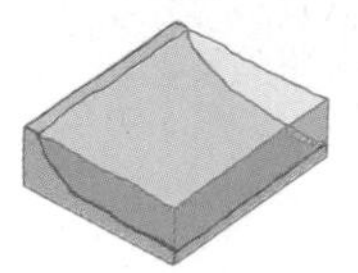

Classification Phylum Mollusca, Class Cephalopoda, Subclass Ammonoidea, Family Macroscaphitidae.
Description The shell is coiled in a flat spiral for the first whorls, but uncoiled at the end without septa. The uncoiled section is marked initially straight and then hooklike, the oral aperture turned back toward the planospiral part. The suture is extremely complex, as in all Lithoceratina, the suborder to which this genus belongs. The coiled part is ornamented with many obvious regular costae; the costae on the uncoiled part near the siphuncle are coarser and further apart, but have no ventral break.
Stratigraphic position and geographical distribution The genus ranges from the Barremian to the Aptian (118–110 million years ago) and is abundant in sediments in central and northern Europe, the Alps and North Africa. The photograph shows the type species *M. yvani* from the Italian Barremian at Val Gardena; it measures 5 cm (2 in).
Note In the spirally coiled whorls, the shell of *Macroscaphites* is indistinguishable from that of *Costodiscus*, which belongs to the same family but has no coiled forms. From studies of the shell's shape, allowing for the live animal's specific weight, center of gravity and position of buoyancy center, it seems likely that the partially uncoiled forms of *Macroscaphites* lived with their aperture directed toward the water surface; their coiled part therefore pointed upwards; their uncoiled part toward the sea floor.

---

96 CRIOCERATITES

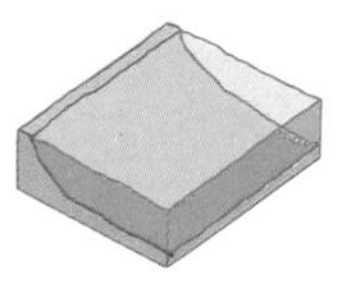

Classification Phylum Mollusca, Class Cephalopoda, Subclass Ammonoida, Family Ancycloceratidae.
Description This is one of the most characteristic known forms of "uncoiled ammonite," so called because there are no points of contact between the successive whorls of the shell. In the case of *Crioceratites* the shell is coiled flat, with no whorls touching. The coil section is subquadrate or oval, and the shell is laterally compressed. The ornament is composed of numerous radial ribs (running transversally to the direction of growth), either straight or gently curved, with a rounded section. At regular intervals, generally every four or five ribs, there is a thicker one equipped with numerous spiny tubercles. The suture is ammonitic, with a highly developed lateral lobe.
Stratigraphic position and geographical distribution The genus is restricted to the Hauterivian and Barremian stages of the lower part of the Cretaceous (125–112 million years ago). It has a broad geographical distribution, occurring in various European localities, as well as in Turkey, Madagascar, Japan, the state of California and Mexico. The example in the photograph comes from the classic locality at Barrême (France); it measures 14 cm (5.5 in).
Note Some species of this genus are used as guide fossils for strata of the Lower Cretaceous. It is believed that the animal was benthonic, meaning that it lived on the sea floor, in this case in cold, shallow waters.

97 ARNIOCERAS

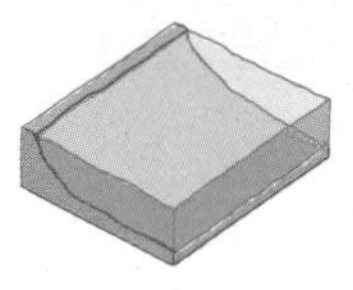

Classification Phylum Mollusca, Class Cephalopoda, Subclass Ammonoidea, Family Arietitidae.
Description A small, flat-coiling shell. It is an evolute form, in the sense that in the course of growth the whorls tend to cover the preceding ones only to a small extent, and the umbilical region is consequently broad and shallow. The sides of the shell are flattened and converge slightly toward the center in the outermost region. The central axis is characterized by the presence of a keel, but unlike many other forms of the same family it lacks two grooves at the sides of this keel. The ornament is provided by regularly-spaced costae, straight for most of their length and then curving forward on the ventral region before dying out. The suture line is ammonitic and marked by a relatively uncomplicated design, with a simple lateral lobe and little fluting.
Stratigraphic position and geographical distribution The genus is limited to the Sinemurian (Lower Jurassic, 190 million years ago). Its geographical distribution, on the other hand, is vast: it occurs in Europe, North Africa, Indonesia, New Caledonia, North America (Alaska and Canada), Mexico and South America (Colombia, Ecuador, Chile, Peru). The photograph shows a slab with a number of specimens (average length 1.5 cm [about ½ in]) from the English Sinemurian.
Note Because of its brief timespan and its broad geographical distribution, *Arnioceras* provides good guide fossils for the Sinemurian stage.

---

98 AMALTHEUS

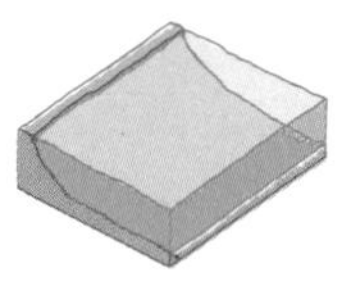

Classification Phylum Mollusca, Class Cephalopoda, Subclass Ammonoidea, Family Amaltheidae.
Description A characteristic form with oxycone shell, meaning that it is laterally compressed, high and has a sharp venter. It is moderately involute, with the last whorl partially covering the preceding one; the umbilical region is therefore moderately developed. The ornament is composed of fairly pronounced ribs and occasionally lateral tubercles in the form of spines. The venter possesses a typical keel, well defined and indented, in the shape of a twisted rope. The suture is ammonitic, of complex design, with two main lateral lobes, the first of which is very wide and long. It is often found in association with aptychi of the *Anaptychus* type, composed of a single valve with a surface covered in thin, concentric striae.
Stratigraphic position and geographical distribution The genus *Amaltheus* is restricted to a zone of the Domerian substage (Upper Pliensbachian, Lower Jurassic, 180 million years ago). It has a very broad geographical distribution, which makes it a very useful guide fossil. It occurs in Europe, North Africa, the U.S.S.R. (Siberia, the Caucasus) and North America (Canada, Alaska, Oregon). The example in the photograph comes from Germany and belongs to the species *A. margarinatus*; it measures 4 cm (1½ in).
Note Some examples have shells still intact, with traces of the original color. These traces are brown, longitudinal striae; the rest of the shell is white.

99 DACTYLIOCERAS

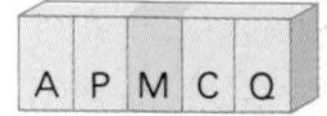

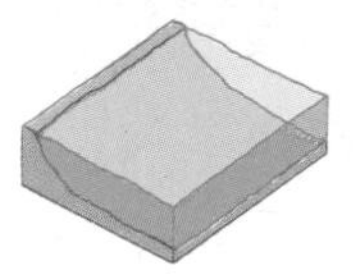

Classification Phylum Mollusca, Class Cephalopoda, Subclass Ammonoidea, Family Dactylioceratidae.
Description One of the most characteristic forms of Lower Jurassic ammonite; its evolute shell has a wide "serpenticone" umbilicus, whose outer whorls only slightly overlap the inner ones. It is ornamented with pronounced closely packed radial ribs, regularly spaced, denser on the inner whorls. The ribs, straight at the sides, are bifurcate on the venter, over which they extend transversally; the ventral region has no keels or grooves. The ammonitic suture has a pronounced lateral lobe.
Stratigraphic position and geographical distribution The genus is typical of the lower part of the Toarcian (end of the Lower Jurassic, 172 million years ago). Its broad geographical distribution makes it an excellent guide fossil for the Lower Toarcian in the form of *D. commune*. It is found in Europe (England, France, Germany, Italy), in North Africa, Iran, Baluchistan, Indonesia, Siberia, Greenland, Alaska, Canada and South America. The example, *D. commune*, from the German Toarcian, measures 4.5 cm (1¾ in).
Note Very abundant in European strata, the forms of the genus *Dactylioceras* have been known for many years. During the Middle Ages at Whitby in Yorkshire (England), where there are very typical beds containing *Dactyloceras*, the inhabitants used to carve eyes and mouths on the ends of these ammonites and then claim that they were snakes turned to stone by St. Hilda, who was venerated in that area.

---

100 HARPOCERAS

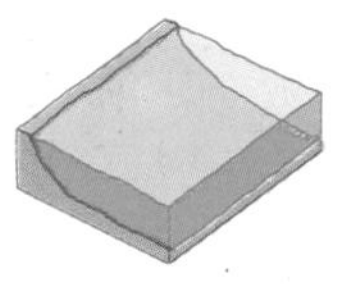

Classification Phylum Mollusca, Class Cephalopoda, Subclass Ammonoidea, Family Hildoceratidae.
Description The flat-coiling shell of *Harpoceras* is very flat in comparison with other forms of the same family. It is very involute, with every whorl almost entirely covering the preceding one, producing a very narrow umbilical aperture at the center of the shell; the umbilical margin is nonetheless sharp, so the umbilicus is well defined. There is a keel on the venter by the siphuncle, but the shell lacks grooves at the side of the keel itself. The ornamentation consists of regular, closely-spaced, curved ribs, concave toward the shell's aperture; these ribs are thicker on the outer half of the shell's sides.
Stratigraphic position and geographical distribution The genus *Harpoceras* is restricted to the lower part of the Toarcian (Lower Jurassic, 178 million years ago). It is abundant in many parts of the world, occurring in Europe (England, France, Germany, Italy), North Africa, the Caucasus, Japan, Indonesia, Canada, the U.S. (Oregon, Nevada) and South America. The example in the photograph, of the species *H. exaratum*, comes from the Italian Toarcian at Furlo in the Apennines; it measures 5.5 cm (2¼ in).
Note Because it was a genus of short duration, with a broad geographical distribution, *Harpoceras* provides certain guide fossils for Toarcian stratigraphy, such as the species *H. falcifer*. The shape of the shell leads paleontologists to believe that the forms included in this genus were well adapted to active swimming.

101 POLYPECTUS

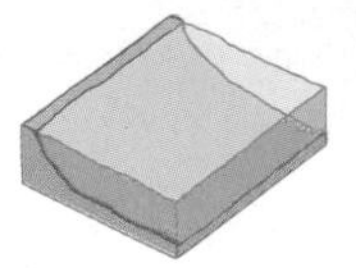

Classification Phylum Mollusca, Class Cephalopoda, Subclass Ammonoidea, Family Hildoceratidae.
Description The shell is typically oxycone in shape, very flattened laterally, but developed in height, with an acute venter whose overall shape is discoidal. It is highly involute, with a narrow umbilical area. The ornament is not accentuated and is formed by large numbers of faint, very closely packed, falciform ribs. The suture is ammonitic, somewhat complex, with a very pronounced lateral and ventral lobe and numerous subsidiary lobes.
Stratigraphic position and geographical distribution The genus, uncommon in fossil form, is restricted to the Upper Toarcian (the upper part of the Lower Jurassic, 168 million years ago). Its geographical spread is limited for the most part to Europe and North Africa. The example in the photograph, which belongs to the species *P. pluricostatus*, comes from the Toarcian of the central Apennines (Italy); it measures 5 cm (2 in).
Note The very acute edge of *Polypectus*, which gives the shell a considerable capacity for cutting through water, indicates that this was a nektonic form, with well-developed swimming abilities.

---

102 PARONICERAS

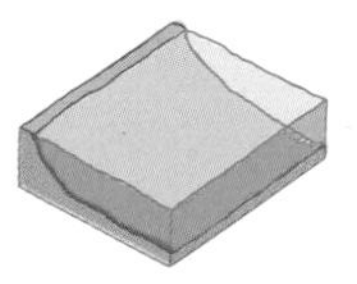

Classification Phylum Mollusca, Class Cephalopoda, Subclass Ammonoidea, Family Hildoceratidae.
Description This genus has a rather unusual shape, considering the family to which it belongs, but it is one that often recurs in genera of ammonites from a variety of different timespans. The shell is stout in appearance, with the coiled section almost as high as it is wide. It is a fairly involute form, with the last whorl covering about two-thirds of the preceding one. The umbilicus is deep and clearly demarcated, with a sharp or rounded margin. In the majority of cases the ornament is either nonexistent or faint, with thin, weak ribs on the sides of the venter; sometimes it takes the form of broad, poorly defined folds on the sides. The venter has no keels or grooves and is mainly rounded, being acute in only one species. The suture is very simple, with insignificant frilled lobes and saddles reminiscent of the ceratitic type, as can be clearly seen in the photograph.
Stratigraphic position and geographical distribution Stratigraphically *Paroniceras* is limited to the Upper Toarcian (the last part of the Lower Jurassic, 168 million years ago). Its geographical distribution is also restricted—the Mediterranean and North Africa. The photograph shows the type species *P. sternale*; it measures 2.5 cm (1 in).
Note The locally abundant *P. sternale* has been used as a guide fossil for the Upper Toarcian, though it is of limited usefulness for this purpose.

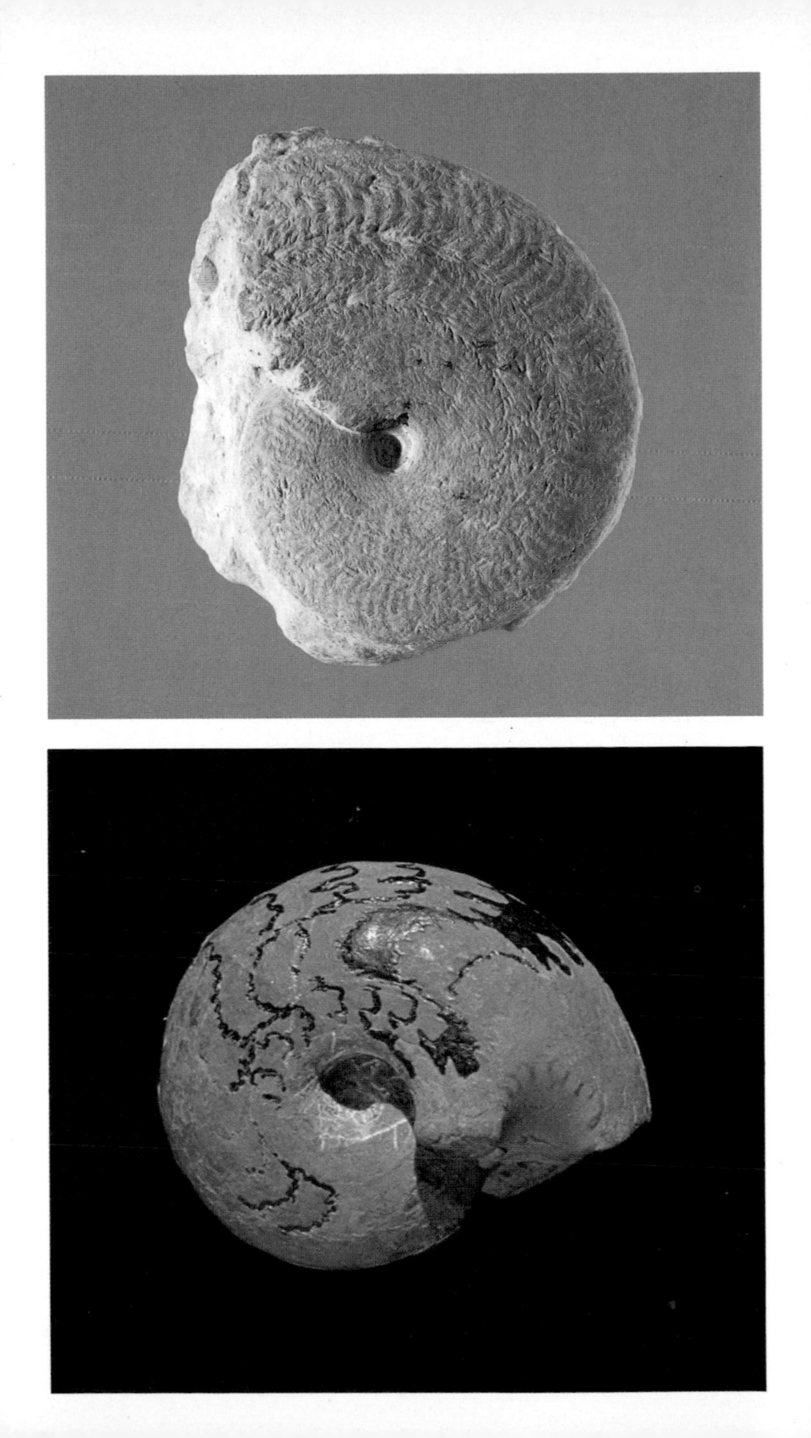

103 OXYPARONICERAS

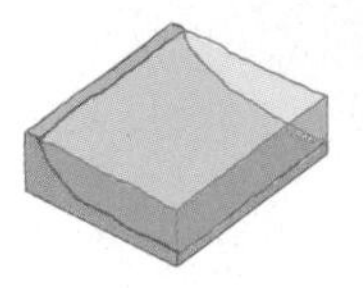

Classification Phylum Mollusca, Class Cephalopoda, Subclass Ammonoidea, Family Hildoceratidae.
Description The shell is either ogival in section, with distended edges, or somewhat flatter, with less curving edges, but with a higher coiled section. The overlap of the preceding whorl by the last one is fairly marked. Ornament is absent and the shell is smooth. There is, however, a pronounced keel, laterally compressed and acute. The suture is very simple compared with those normally found in Jurassic ammonites and is of a type very similar to the ceratitic; in the photograph this is artificially emphasized.
Stratigraphic position and geographical distribution The stratigraphic position of the genus is limited to the Upper Toarcian (the last part of the Lower Jurassic, 168 million years ago). Its geographical distribution is also restricted: only a few examples, generally a few centimeters in diameter, have been found in France, Spain and Italy. The example in the photograph comes from Furlo in the Apennines (Italy) and belongs to the species *O. buckmanni*; it measures 6.2 cm (about 2½ in). In the photograph, the design of the sutures has been highlighted in Indian ink.
Note Together with *Paroniceras*, from which it appears to be derived, this genus belongs to a small group of ammonites which, because of their shell shape, their suture and their distribution, are unique among Lower Jurassic ammonites.

---

104 LEIOCERAS

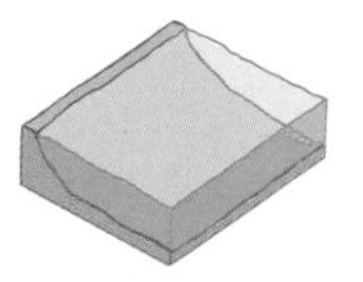

Classification Phylum Mollusca, Class Cephalopoda, Subclass Ammonoidea, Family Graphoceratidae.
Description This genus is characterized by a very flattened and involute planispiral shell, of typically platyoxycone form, meaning that it is discoidal with a sharp external (ventral) edge; the coil section is ogival. The shape is involute, with the outermost whorl almost completely covering the innermost ones, and with a rapid upward growth of the coil. There is a sharp keel on the ventral margin, but it is not bordered by furrows. The shell is ornamented with regular, infinitesimally fine, sickle-shaped ribs that bifurcate from the inner third of the whorl. The suture is ammonitic, not overly intricate, with well-defined, but not highly frilled saddles and lobes.
Stratigraphic position and geographical distribution The timespan of the genus is limited to the Aalenian (Middle Jurassic, 166 million years ago). It is known in western Europe, North Africa, Anatolia, the Caucasus and Iran. The example in the photograph comes from the German Aalenian and belongs to the type species *L. opalinum*; it measures 3.7 cm (about 1½ in).
Note Given its relative abundance in layers and its brief timespan, the species *L. opalinum* is a typical guide fossil for the Aalenian in western Europe. The genus is one of the earliest representatives of the family Graphoceratidae, which, with its compressed, involute forms, was one of the most characteristic ammonite families of the beginning of the Middle Jurassic.

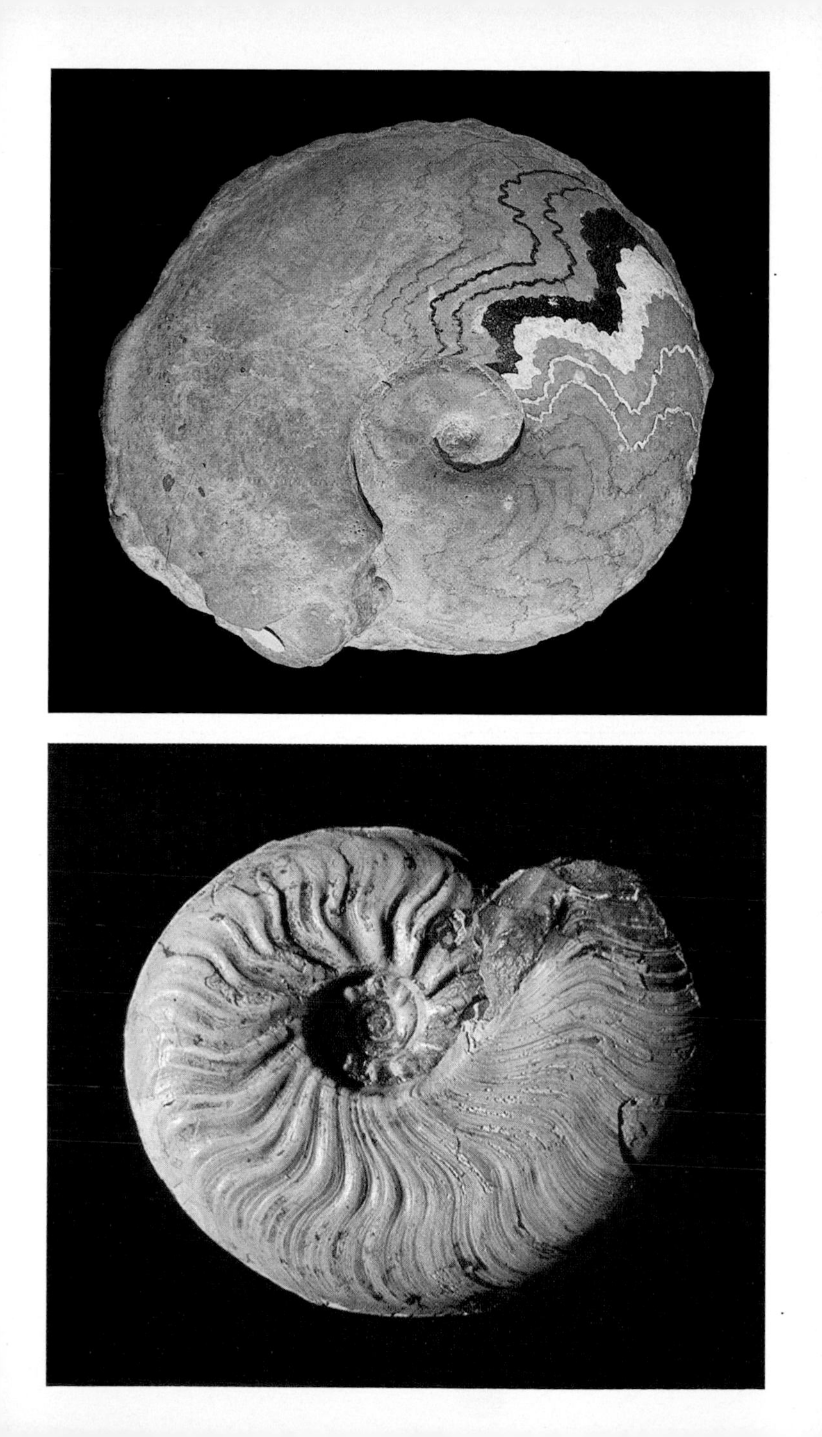

105 PHYMATOCERAS

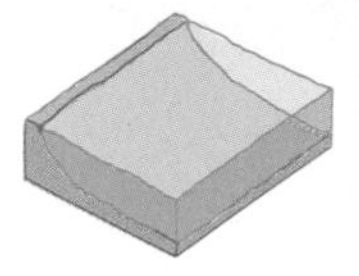

Classification Phylum Mollusca, Class Cephalopoda, Subclass Ammonoidea, Family Phymatoceratidae.
Description This genus comprises a group of forms possessing for the most part evolute shells, with the last whorl covering only a third of the preceding one. The umbilicus is wide and usually shallow. Seen in section it is more or less triangular, with a height greater than its width. The ornament, which is very well developed and variable, sometimes consists of pronounced ribs with a more or less markedly sinusoidal arrangement. The ribs may either be single or they may bifurcate or trifurcate in correspondence with prominent tubercles situated on the innermost part of the side, almost on the umbilical margin. In almost all the forms of the genus the venter is usually equipped with a keel bordered by furrows. The suture line is ammonitic, with pronounced saddles and lobes.
Stratigraphic position and geographical distribution The genus is restricted to the Upper Toarcian (the last part of the Lower Jurassic, 168 million years ago). Its geographical distribution, by contrast, is broad: examples of *Phymatoceras* are found in Europe, mainly in countries of the Mediterranean basin, in North Africa, Anatolia, Japan, Alaska and South America. The example in the photograph measures 8.2 cm (about 3¼ in).
Note The species *P. erbaense* is used as a zone indicator in the Mediterranean Toarcian.

---

106 HAMMATOCERAS

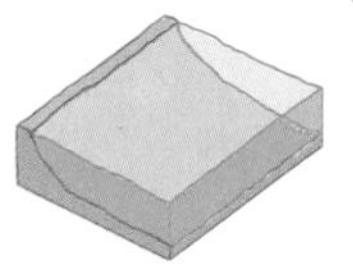

Classification Phylum Mollusca, Class Cephalopoda, Subclass Ammonoidea, Family Hammatoceratidae.
Description Fossil forms with flat coiling shells, whose degree of involution can vary considerably: there are fairly evolute forms with whorls partially covered by the preceding one and more involute ones with a greater degree of overlap. The shape of the coiled section may also vary as a consequence: as well as forms with a more or less triangular section, there are those with an almost trapezoidal one. The ornament is generally provided by pronounced ribs, of varying thickness, which can be either single or bifurcated and corresponding to tubercles sited on the inner third of the sides, close to the umbilical margin. The venter, lacking furrows, generally has a rather indistinct keel where the ribs end. The suture line is of the ammonitic type, with a design made fairly elaborate by various saddles and lobes.
Stratigraphic position and geographical distribution The genus is limited to the Upper Toarcian and the Aalenian (between the Lower and Middle Jurassic, 168 million years ago). Its geographical distribution, however, is extensive: it occurs in Europe, North Africa, Indonesia, Canada and South America. The example in the photograph, of the species *H. perplanum*, comes from the Italian Apennines and measures 11.3 cm (about 4½ in).
Note Because of its short timespan and its relative abundance in many European layers, the species *H. insigne* is used as a guide fossil for the Upper Toarcian.

107 HAPLOCERAS

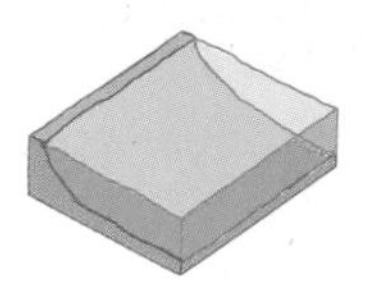

Classification Phylum Mollusca, Class Cephalopoda, Subclass Ammonoidea, Family Haploceratidae.
Description Medium-sized shell, with a flat coil and with the last whorl mostly covering the preceding one. The umbilicus is therefore fairly narrow and deep, with rounded umbilical margins. In section it is more or less rectangular, with its height greater than its width. The venter is rounded, lacking both a keel and furrows. Ornamentation is absent or at best present in the form of slight folds on the ventral region of the body chamber (it should be recalled that the body chamber, the last space in the shell to be inhabited by the living organism, is characterized, in fossils, by the absence of sutures that divide the rest of the shell into a series of chambers). The suture line is ammonitic, with deeply incised and frilled saddles and lobes.
Stratigraphic position and geographical distribution The genus is typical of the final stages of the Jurassic period (Kimmeridgian and Tithonian, 140–100 million years ago). Its geographical distribution is extensive, with examples of *Haploceras* being found in Europe, North Africa, Tanzania, Asia, the state of Texas, Mexico and Cuba. The example in the photograph belongs to the type species *H. elimatum* and comes from the Kimmeridgian in the Italian Apennines; it measures 6.7 cm (about 2½ in).
Note Ammonites of the genus *Haploceras* are found in association with double-valved aptychi such as *Lamellaptychus*.

---

108 PHANEROSTEPHANUS

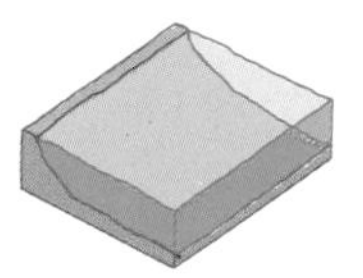

Classification Phylum Mollusca, Class Cephalopoda, Subclass Ammonoidea, Family Perisphinctidae.
Description An ammonite, sometimes of considerable size (a few dozen centimeters), with a flat, fairly evolute coil in which the last whorl only partially covers the preceding one. The height of the whorl section is greater than its width. The innermost whorls have single ribs on the inner third of their edge, giving rise to triple ribs toward the middle. The ribs traverse the rounded venter, which has neither keel nor furrows. The outermost whorls have no such triple ribs, ornament being provided by thick, single ribs that give way to large tubercles on the umbilical region. The suture is ammonitic.
Stratigraphic position and geographical distribution The stratigraphic position of the genus is limited to the Tithonian, the final stage of the Jurassic period, 140 million years ago. Its geographical distribution is also limited: examples of the genus occur in Kurdistan and in Europe in very small numbers. The example in the photograph comes from the Italian Apennines and measures 13.5 cm (about 5 in).
Note The genus forms part of a large group of Perisphinctidae that were very widespread in numerous different forms in the seas of the Upper Jurassic.

109 MACROCEPHALITES

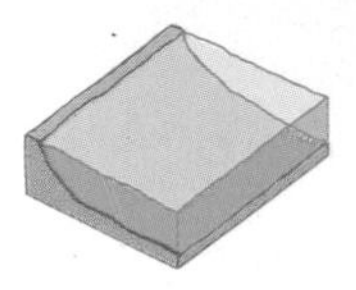

Classification Phylum Mollusca, Class Cephalopoda, Subclass Ammonoidea, Family Macrocephalitidae.
Description An ammonite that can achieve a considerable size, with a fat shell and a very rounded venter. The shell is also highly involute, the last whorl almost completely covering the preceding whorls; the umbilicus is consequently deep and narrow. The body chamber, recognizable as the part of the shell coming after the last suture line, is smooth, whereas the remainder of the shell is densely decorated with straight, forward-pointing ribs that cross the venter, which has no keel or furrows. The suture line, of the ammonitic type, is fairly intricate, with a well-developed ventral lobe and the second lateral lobe possessing a single very thin and elongate central lobe.
Stratigraphic position and geographical distribution The genus is confined to the Lower Callovian (Middle Jurassic, 162 million years ago). This limited timespan is counterbalanced by a very broad geographical distribution: the genus is found in Europe, northern and central-eastern Africa, the Caucasus, Indonesia, New Guinea and North and South America. The example in the photograph comes from the Swiss Callovian at Herznach and measures 6.1 cm (about 2½ in).
Note The genus originated in the ancient Tethys ocean, located between Asia and Europe to the north and Africa to the south; because of its broad distribution the species *M. macrocephalus* is a guide fossil for the Lower Callovian.

---

110 PARKINSONIA

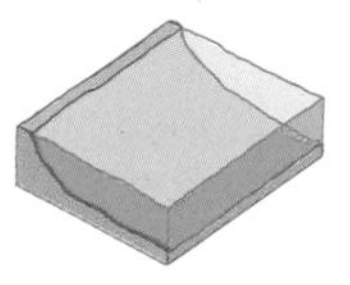

Classification Phylum Mollusca, Class Cephalopoda, Subclass Ammonoidea, Family Parkinsoniidae.
Description An evolute, flat coiling shell, with the last whorl only partially covering the preceding ones; it is also laterally compressed, with a subtrapezoidal whorl section whose height is greater than its width. The ornament comprises thick, sharply outlined ribs, which start on the umbilical margin and trifurcate close to the venter, broken only by a faint keel at the siphuncle; the ribs are slightly forward curving from the inner third of the whorls and may sometimes bear lateral tubercles. The shell occasionally reaches a considerable size. The suture, of ammonitic type, is not excessively complicated.
Stratigraphic position and geographical distribution The genus is restricted to the Upper Bajocian (Middle Jurassic, 175 million years ago). It is found in Europe, where it is typical of the English Bajocian; in North Africa, certain areas of the U.S.S.R. (Crimea, Azerbaijan, the Caucasus) and in Iran. The example in the photograph is the species *P. parkinsoni*, comes from France and measures 7.5 cm (3 in).
Note The genus *Parkinsonia,* together with other genera of the family, is one of the earliest representatives of the Perisphinctacea, a "superfamily" of ammonites that enjoyed wide distribution in the seas of the Middle and Upper Jurassic.

111 ASPIDOCERAS

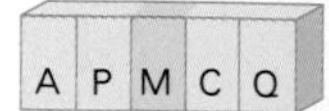

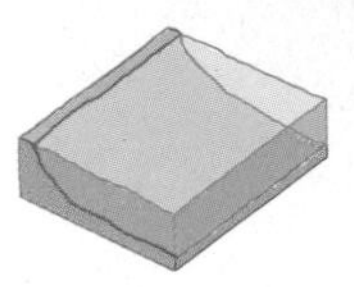

Classification Phylum Mollusca, Class Cephalopoda, Subclass Ammonoidea, Family Aspidoceratidae.
Description A typical form of Jurassic ammonite, with a shell coiling in a plane spiral, moderately involute, with partial overlap of the preceding whorls by the outer whorl. The shell's cross section is depressed or square, with multiple rounded angles. The ornament comprises two series of lateral tubercles sited around the middle of the spire; sometimes the outermost line of tubercles grows fainter and occasionally even disappears. Some species have a venter ornamented with ribs. The suture line, of ammonitic type, is generally simple, with a broad ventral lobe and lateral lobes with only one point.
Stratigraphic position and geographical distribution The genus *Aspidoceras* is typical of the Upper Jurassic, especially the Kimmeridgian stage (150 million years ago), and enjoys a broad geographical distribution: it is found in Europe, northern and central-eastern Africa, the U.S.S.R., Japan, New Zealand, Central America and Argentina. The example in the photograph comes from the Italian Kimmeridgian in the Apennines and belongs to the species *A. acanthicum*; it measures 6 cm (2½ in) in diameter.
Note Examples of the genus *Aspidoceras* are usually discovered associated with very thick aptychi of the type *Laevaptychus*.

---

112 POLYPTYCHITES

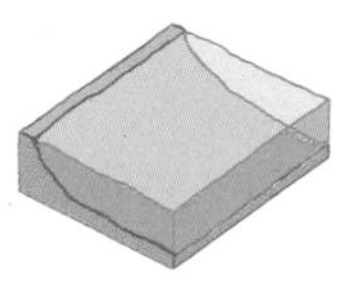

Classification Phylum Mollusca, Class Cephalopoda, Subclass Ammonoidea, Family Olcostephanidae.
Description Shell coiling in a plane spiral whose degree of involution varies from moderately involute to highly evolute; the width of the umbilicus also varies as a consequence. The whorl section is depressed, greater in height than in width, with the widest part being the inner third. The venter is round, with no furrows or keel. The ornament is very marked and consists of strong, elongate tubercles spreading out from the umbilical margin; these also give rise to two or more ribs, which may themselves bifurcate or trifurcate at the edges. The innermost whorls of some forms possess high, well-spaced ribs that bifurcate close to the umbilical margin. The suture is ammonitic, with a broad lateral lobe.
Stratigraphic position and geographical distribution The genus occurs from the Lower Valangian to the Upper Valangian (Lower Cretaceous, 130–125 million years ago). It is found in northern Europe, northern Asia, Mexico and the state of California. The example in the photograph comes from the German Valangian and measures 6.8 cm (about 2½ in).
Note *Polyptychites* is a typical genus of the northern fauna that characterized the arctic and northern regions of Europe, America and Asia during the Mesozoic.

15708

113 LAMELLAPTYCHUS

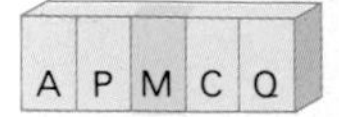

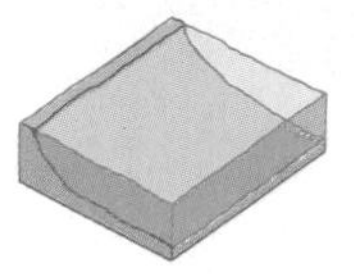

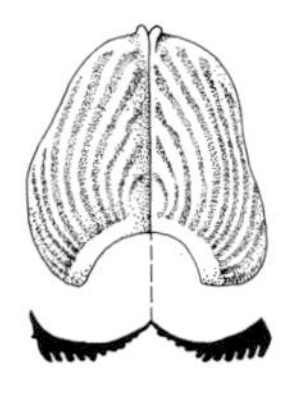

Classification Phylum Mollusca, Class Cephalopoda, Subclass Ammonoidea.
Description This genus embraces aptychi whose typical structure is composed of a pair of elongate calcareous valves, subtriangular in shape, ornamented on the outer surface with gently curving and clearly marked longitudinal ribs. These aptychi consist of a thick calcareous outer layer, made up of several superimposed layers, and a thin, horny inner layer.
Stratigraphic position and geographical distribution Aptychi of the genus *Lamellaptychus* are found in Jurassic sediments; in particular, they are found typically associated with ammonites of the family Oppelidae, characteristic of the Upper Jurassic (165–140 million years ago). The example in the photograph comes from the Italian Upper Jurassic at Suello in Lombardy, it measures 4 cm (1½ in).
Note Structures of this type may comprise calcareous valves (aptychi) or a single chitinous valve (anaptychi). Anaptychi are first found in the Devonian, associated with the first known ammonites, and reached the period of their greatest expansion from the Upper Paleozoic to the Lower Jurassic. Aptychi do not appear until later, in Jurassic sediments, and occur up until the Cretaceous, at the end of which they disappeared when ammonites became extinct. Anaptychi have recently been interpreted as being the lower jaws of ammonites.

---

114 LAEVAPTYCHUS

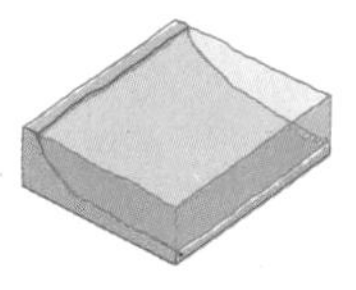

Classification Phylum Mollusca, Class Cephalopoda, Subclass Ammonoidea.
Description Aptychi such as *Laevaptychus* consist of paired, symmetrical, calcareous valves, frequently occurring on their own in sediments or contained within the body chamber of ammonite shells. From a morphological point of view they recall the shells of bivalves or other organisms, an observation that led to their misleading interpretation in the past. Aptychi of the *Laevaptychus* type are very frequent in Jurassic rocks. Less thick—sometimes very thin—in their outer, calcareous layer, they possess an inner layer composed of organic matter; the calcareous part is made up of small calcite tubes of such density that, in a thin section, they give the impression of spongy tissue.
Stratigraphic position and geographical distribution Aptychi of the genus *Laevaptychus* range from Middle to the Upper Jurassic (170–140 million years ago) and belong to ammonites of the family Aspidoceratidae. The example in the photograph, which is large in size (20 cm or 8 in), comes from Tierra del Fuego, Argentina.
Note Aptychi were for a long time interpreted as being the opercula of ammonites, due to the fact that they are often discovered in the shells of these animals in a position appropriate to this function. In recent years, however, specialists have come to regard aptychi as forming part of the masticatory apparatus of these extinct cephalopods.

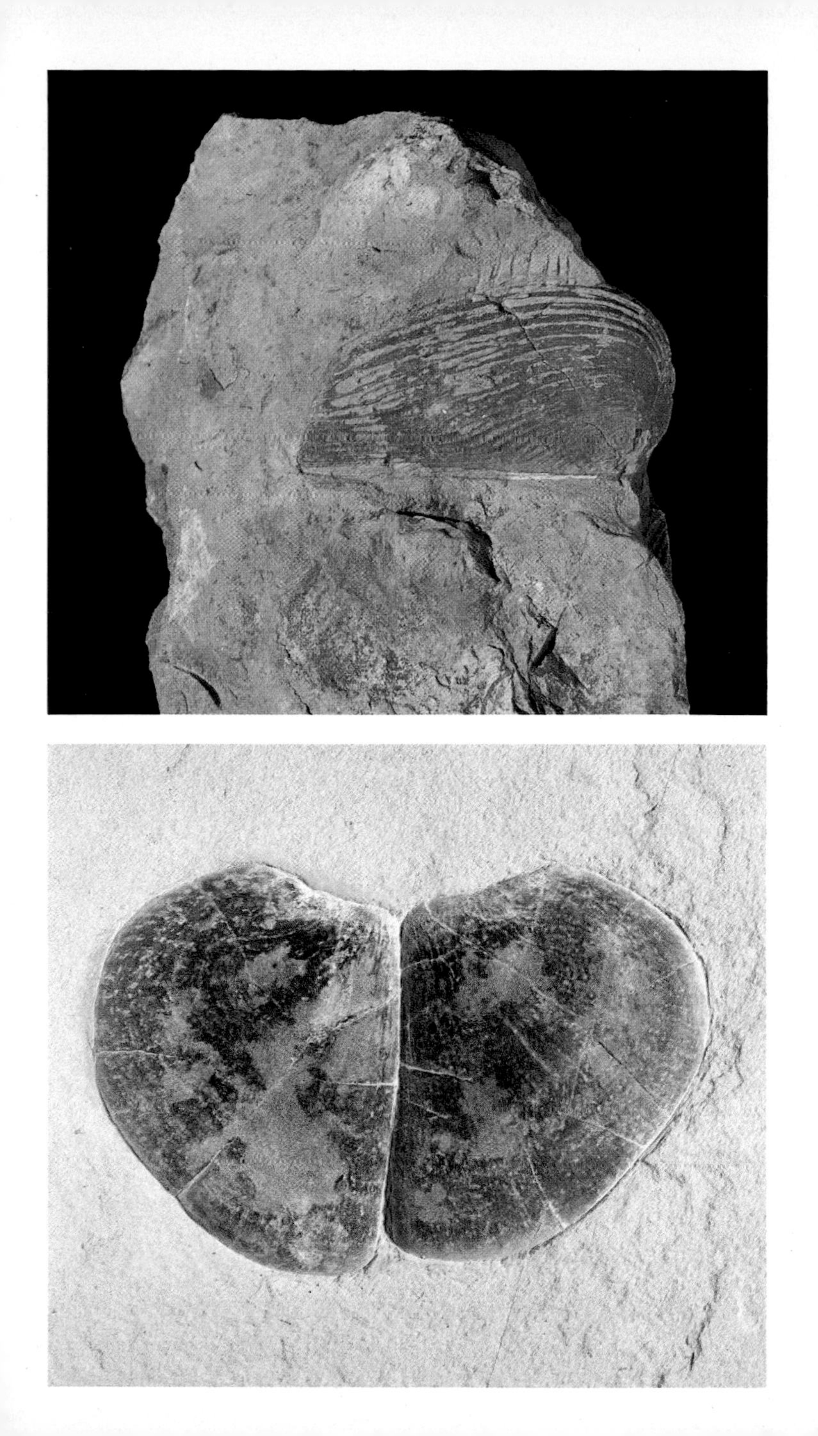

115 BELEMNOPSIS

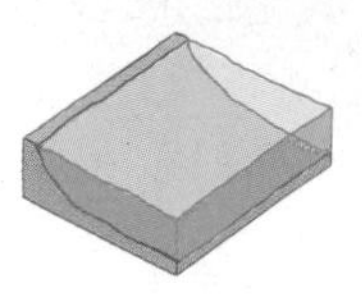

Classification Phylum Mollusca, Class Cephalopoda, Subclass Belemnoidea, Family Belemnitidae.
Description Extinct cephalopods whose body was equipped with an internal shell. The shells of belemnites, which except for rare cases are the sole part to survive in fossil form, consist of a chambered section or "phragmacone," which because of its fragility is often poorly preserved or even absent in fossils, and a very solid guard, which is the section that usually survives. This guard is normally of elongate form, very solid and tapering at the posterior end, which has a pointed tip that can also be rounded in some genera, and is made of calcite. The function of such a structure must have been to act as a counterbalance to the lighter, anterior end of the body and insure greater stability while swimming.
Stratigraphic position and geographical distribution The genus is found in sediments of the Lower Jurassic (190 million years ago) in Europe and Asia. The slab in the photograph comes from England; the examples contained in it measure 6–7 cm (2½–2¾ in).
Note Belemnites are often widespread in rocks of the Mesozoic era, although the earliest representatives are known from the Upper Carboniferous and a few species survived up until the Eocene. They must have been similar in appearance to the modern cuttlefish, possessing ten tentacles equipped with a double row of horny hooks and an ink sac. Like modern cuttlefish, belemnites must have been organisms capable of active swimming, which fed on fish or crustacea.

---

116 RHOMBOTEUTHIS

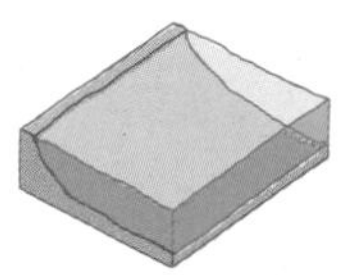

Classification Phylum Mollusca, Class Cephalopoda, Subclass Dibranchiata, Order Teuthoidea, Family Plesioteuthidae.
Description A primitive teuthoid, related to the modern cuttlefish and squid, whose general morphology it recalls. Small, up to approximately 9 cm (3½ in) long and 1.5 cm (½–¾ in) wide, the outer surface of the body displays longitudinal ribs, like certain modern Teuthoidea. The posterior extremity is pointed, with two long narrow natatory fins. The internal shell, almost as long as the animal's body, is narrow and little calcified. There are two large eyes and also ten arms, visible in the anterior region; the sessile arms carry a double row of well-developed suckers.
Stratigraphic position and geographical distribution The genus is known from the species *R. lehmani* in the French Callovian (Middle Jurassic, 162 million years ago) at La Voulte-sur-Rhône in the Ardèche, from which the example in the photograph also comes.
Note Like the one in the photograph, examples of this genus sometimes occur in an exceptional state of preservation in the French sediments, complete with their soft parts in three dimensions. This fine and highly unusual degree of preservation is due to the rapid burial of the dead animals and to the impregnation of their soft tissues with pyrites. The sea floor must have been so poor in oxygen that it denied life to those scavenging organisms that fed on the carrion of others, thereby permitting the fossilization of entire remains.

117 AYSHEAIA

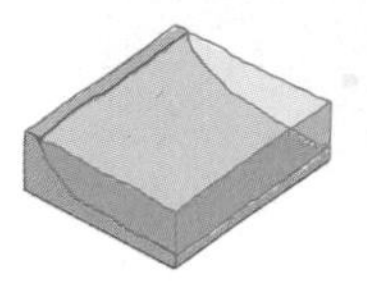

Classification Phylum Arthropoda, Class Onychophora, Order Protonychophora, Family Aysheaiidae.
Description A rare arthropod, known only from a few specimens, ranging in length from 1 cm (½ in) to 6 cm (2½ in). Its cylindrical body has a pair of appendages on the head and ten pairs of stocky, elongate limbs. The head is distinct from the body, while the terminal region fuses with the base of the last pair of limbs. Fossils show that the body was covered in a flexible, waxy layer subdivided into at least 12 successive, annulate segments. These rise up on the dorsal region to form an almost crestlike ridge and are adorned with tubercles. The mouth is also visible, surrounded by papillae. The limbs are more or less conical; they too have 12 annulate segments, which may have spines pointing either forward or backwards. The limbs possess claws.
Stratigraphic position and geographical distribution The genus contains a single species, *A. peduncolata*, hitherto found solely in the Burgess Shale formation in British Columbia (Canada). This dates from the Middle Cambrian (530 million years ago) and is famed for the exceptional way it has preserved even soft-bodied organisms in fossil form.
Note *Aysheaia* has been attributed to the class Onychophora, to which the modern *Peripatus*, a New Zealand living fossil, also belongs. The animal lived among colonies of sponges and it is possible that it fed on their soft parts.

---

118 MARRELLA

Classification Phylum Arthropoda, Class Trilobitoidea, Order Marrellida, Family Marrellidae.
Description A small, trilobitomorphic arthropod with a maximum length of 2 cm (¾ in). It possesses a small, wedge-shaped head shield, from the front of which there emerge two long and thick, forward-curving spines. The body is narrow and of almost circular section. To the side of the cephalon is a segmented antenna. The body is subdivided into segments, each of which carries filamentary appendages. Unlike trilobites, however, there are no lateral expansions of the exoskeleton (pleurae).
Stratigraphic position and geographical distribution The genus has only a single species, *M. splendens*, which occurs in the Burgess Shale in British Columbia (Canada) and dates from the Middle Cambrian (530 million years ago). The example in the photograph measures 1.5 cm (½ in –¾ in).

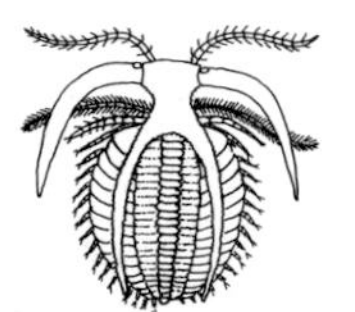

Note This organism, highly enigmatic because of the lack of any well-defined affinity with other groups of arthropods (even though in many respects its general form recalls that of a trilobite), is very widespread in the Burgess Shale; so much so, in fact, that it represents more than a third of the total fauna discovered there. It was an inhabitant of the sea floor, over which it moved by walking on its forelegs, and was probably also able to swim by using its filamentary posterior appendages. Its diet consisted of minute organic particles present in mud on the sea floor.

119 SIDNEYA

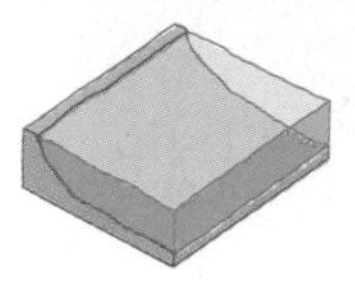

Classification Phylum Arthropoda, Class Trilobitoidea, Order Limulavida, Family Sidneyiidae.
Description An arthropod of uncertain systematic affinities with other groups of living and fossil arthropods. It is a large organism (the example in the photograph is more than 13 cm or 5½ in long), whose body is composed of an anterior part or "cephalon," equipped with a backward-pointing ventral plate covering the mouth; a thorax, made up of nine articulated segments or somites that constitute the broadest part of the animal; a posterior part or "abdomen," which is, by contrast, either very narrow or cylindrical in shape, with an exoskeleton subdivided into two or three segments. At the end of the abdomen is the telson, with two large, fan-shaped appendages at its sides. The eyes are borne on a stalk and there are also two antennae. The thorax possesses appendages adapted to walking.
Stratigraphic position and geographical distribution Like the two preceding genera, the genus *Sidneya* has only one species, *S. inspectans*, found in the Burgess Shale formation in British Columbia (Middle Cambrian, 530 million years ago), well known for its rich fauna of trilobites and primitive arthropods.
Note *Sidneya* must have lived on the sea floor, judging from the considerable development of its ambulatory limbs. It was a carnivorous animal that, judging from the fossilized remains in its stomach, must have fed on ostracod crustaceans, hyolithid molluscs and small trilobites.

---

120 BURGESSIA

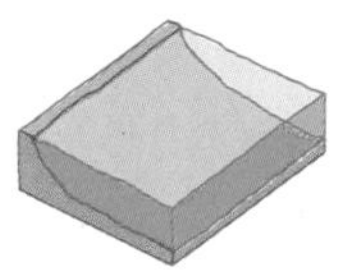

Classification Phylum Arthropoda, Class Trilobitoidea, Order Burgessiida, Family Burgessiidae.
Description An arthropod whose most obvious characteristic is a large carapace with a long posterior spine. The carapace is convex, almost circular in shape, with a diameter of a dozen or so centimeters (5 in). The body is divided into segments, contained entirely below the carapace and composed of a cephalic region and a posterior region or trunk. The cephalic portion of the body bears four pairs of appendages, the first pair of which are segmented antennae, while the others each consist of a segmented leg and a small flagellum. The trunk is composed of eight segments, the first seven of which bear biramous appendages consisting of a leg and a gill, whereas the last segment bears simple, uniramous appendages; on the latter segment there is also a long spine.
Stratigraphic position and geographical distribution The genus has only a single species, *B. bella*, exclusive to the Burgess Shale in British Columbia (Canada), a formation dating from the Middle Cambrian (530 million years ago).
Note The shape of *Burgessia* cannot help but recall that of the living *Limulus*, a xiphosurid arthropod widespread in modern seas: it is probable that *Burgessia*, like *Limulus*, was an inhabitant of the sea floor and used its limbs to move along it; there is less certainty, however, about its ability to swim by moving its gills.

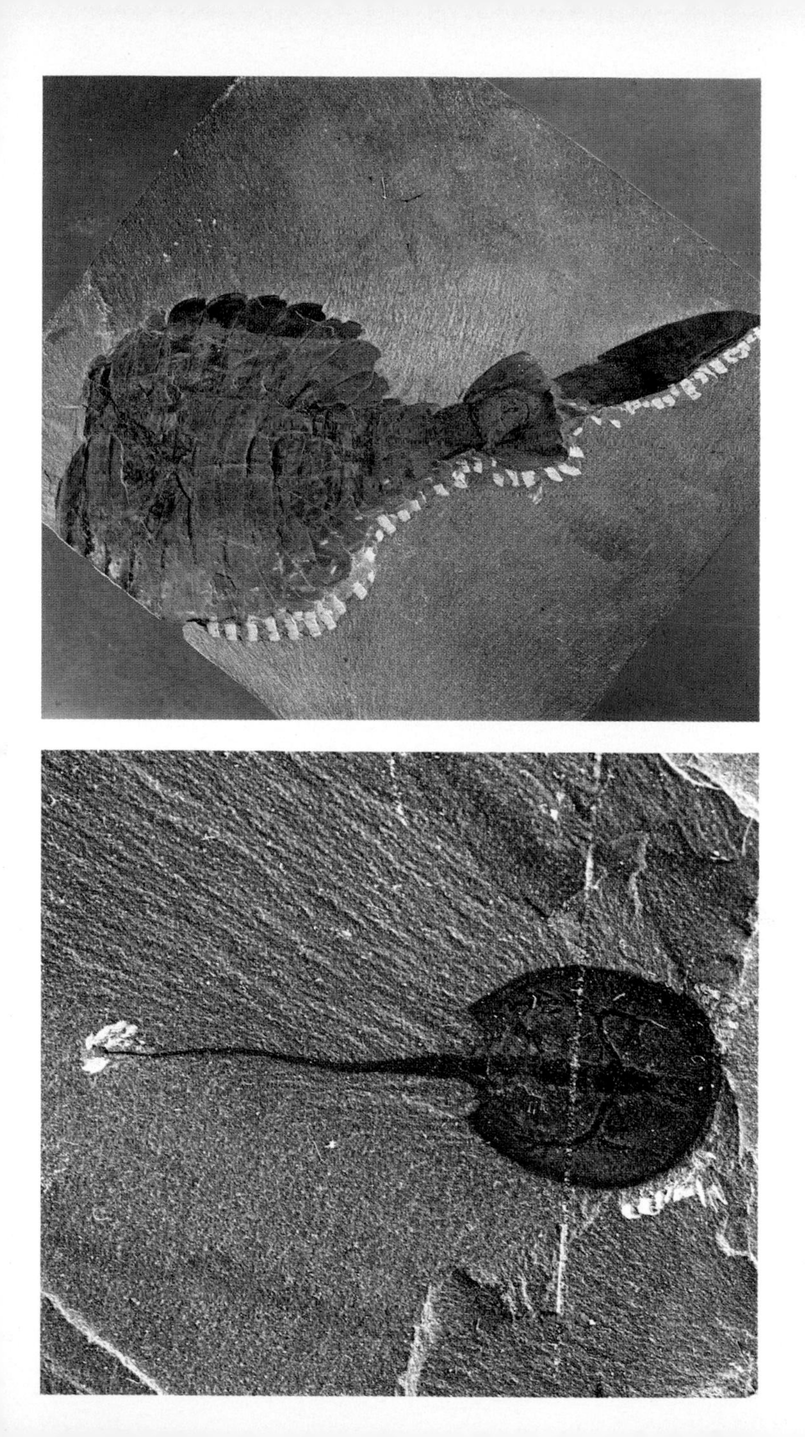

121 OLENELLUS

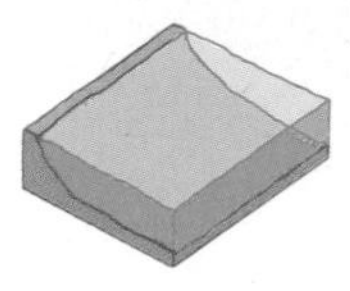

Classification Phylum Arthropoda, Class Trilobita, Order Redlichiida, Family Olenellidae.
Description A trilobite with an exoskeleton of subovoidal form, characterized by a broad cephalon and a small pygidium. The glabella, the raised portion at the center of the cephalon, is elongate and almost cylindrical in shape, with a rounded frontal lobe and three pairs of lateral furrows, and extends as far as the anterior border of the cephalon; the palpebral lobes, above the eyes, are elongate. The genal spines, positioned at the posterior extremities of the cephalon, are short but pronounced. The thorax, as is characteristic of some Olenellidae, is divided into two sections: the anterior part of the thorax (prothorax) contains 14 pleurae, normally developed and ending in more or less sharply retroflex and mainly very long pleurae; the posterior part of the thorax (opisthothorax) contains a variable number of segments (about 30) bearing somewhat rudimentary pleurae, with the first of these posterior segments carrying a long, central, retroflex spine. The pygidium is very small.
Stratigraphic position and geographical distribution This is a very old genus of trilobite, found exclusively in the Lower Cambrian (570–550 million years ago) in North America, Greenland and Scotland. The example, from the Lower Cambrian in Lincoln County, Nevada, has been assigned to the species *O. clarki*; it measures about 8 cm (3¼ in).
Note Olenellidae, a primitive trilobite family, atypically could not enroll.

---

122 PARADOXIDES

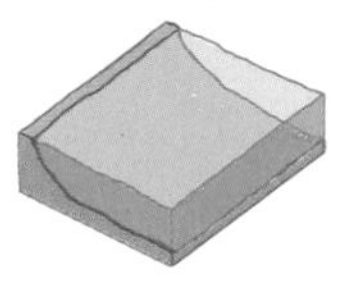

Classification Phylum Arthropoda, Class Trilobita, Order Redlichiida, Family Paradoxidae.
Description Primitive form of trilobite that was capable of achieving considerable dimensions, up to 20 cm (8 in) or more. The cephalon is highly developed, with a globular glabella expanding as far as its anterior margin; it is very broad anteriorly, with the eyes drawn back toward the center, and displays very clearly marked furrows. The cephalon also possesses two genal spines at its rear, and these are long, robust and sharply retroflex. The thorax consists of between 16 and 21 segments, and the pleurae terminate in obliquely retroflex spines whose length increases as they near the end of the body. The two last spines, which are the most robust, stretch straight back beyond the pygidium, itself either small with an entire margin or posteriorly equipped with few pairs of spines.
Stratigraphic position and geographical distribution The genus is known from the Middle Cambrian (550–530 million years ago); it occurs in Europe, North Africa and the eastern U.S. The example in the photograph is from the Spanish Middle Cambrian and measures approximately 8 cm (3¼ in).
Note The presence of this primitive trilobite genus, probably derived from Olenellidae, has come to characterize an ancient faunal province of the Middle Cambrian known as the North Atlantic province, geographically delineated by the geographical distribution of *Paradoxides*.

123 OGYGOPSIS

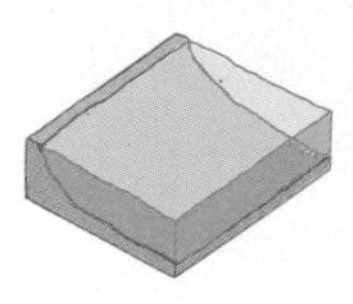

Classification Phylum Arthropoda, Class Trilobita, Order Corynexochida, Family Ogygopsidae.
Description This is a trilobite of a very different form from those seen so far. The exoskeleton is more compact and broad, of the isopygous type (the pygidium is almost identical in size to the cephalon). The cephalon, which possesses two very pronounced, retroflex genal spines, displays a prominent glabella, lacking furrows and extending as far as the anterior margin; the eyes are of medium size and situated roughly halfway along the glabella. The thorax is composed of eight segments, with broad pleurae that have no spines at their extremities. The pygidium is well developed, with a multi-segmented axis stretching as far as the posterior edge; the lateral portions of the pygidium are crossed transversally by pleural furrows; its edge, posteriorly rounded, is generally entire and spineless.
Stratigraphic position and geographical distribution The genus is known from the Middle Cambrian (530 million years ago) of North America. The example in the photograph comes from the Burgess Shale in British Columbia (Canada), and measures approximately 5 cm (2 in).
Note The order Corynexochida includes a large number of genera restricted solely to the Lower and Middle Cambrian, their forms characterized by an elliptical carapace and a broad pygidium.

---

124 MODOCIA

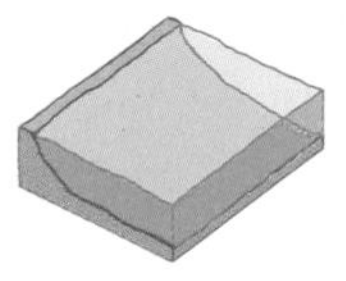

Classification Phylum Arthropoda, Class Trilobita, Order Ptychopariida, Family Marjumiidae.
Description The exoskeleton of this trilobite is elliptical in shape, with a cephalon greater than its pygidium. The glabella, posteriorly wide, narrows toward the front, with a rounded anterior border, and has three pairs of short, faint furrows at the sides. The eyes are unpronounced and much smaller than the eyes of other trilobites. There are genal spines of varying length, but well marked and obliquely retroflex. The thorax, fairly broad at the front, gradually narrows toward the posterior: it is composed of 14 segments, with the spineless pleurae rounded at the tip. The pygidium is rounded and may possess small marginal spines in its anterior section. The surface of the exoskeleton is finely granulated.
Stratigraphic position and geographical distribution The genus ranges from the upper part of the Middle Cambrian to the end of the Cambrian (550–500 million years ago) and is typical of the North American Cambrian. The example in the photograph belongs to the species *M. brevispina* and comes from Liberty, Idaho; it measures approximately 3 cm (1¼ in).
Note The family to which *Modocia* belongs contains numerous genera found mainly in North America; its representatives lived solely from the Middle to the Upper Cambrian.

125 BASIDECHENELLA

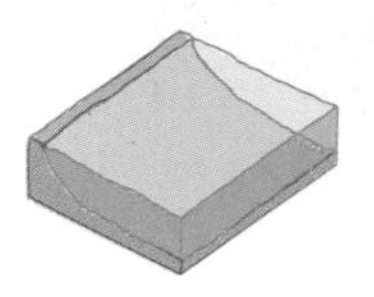

Classification Phylum Arthropoda, Class Trilobita, Order Ptychopariida, Family Proetidae.
Description Medium to small trilobites with an elongate, elliptical exoskeleton and a well-developed pygidium. The anteriorly rounded glabella broadens toward the posterior; it is granulated, with two slightly accentuated lateral furrows on the rear third. The palpebral lobes are clearly marked. The cephalon has obliquely retroflex genal spines. The thorax has 11 segments, with spineless pleurae. The pygidium is fairly elongate; its axis has 12 or 13 rings; the pleural area has about eight ribs and both axis and pleural area are ornamented with tubercles regularly arranged in transversal rows. The pygidium margins are flattened and entire.
Stratigraphic position and geographical distribution *Basidechenella*, regarded by some scholars as a subgenus of *Dechenella*, is restricted to the upper part of the Lower Devonian (380 million years ago) and only doubtfully recorded in the Middle Devonian. It occurs in Europe and North America. The example which measures about 5 cm (2 in), is from the Devonian in Livingston County, New York.
Note The family Proetidae includes numerous genera, whose timespan ranges from the Middle Ordovician to the Lower Carboniferous. Members of this family can enroll in the way most typical of trilobites: the animal rolled itself into a ball, the pygidium curling up to meet the cephalon, protecting the entire lower surface with the dorsal exoskeleton.

---

126 AULACOPLEURA

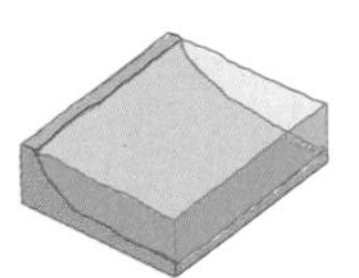

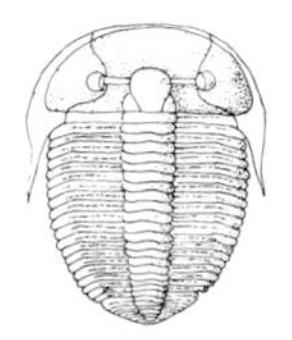

Classification Phylum Arthropoda, Class Trilobita, Order Ptychopariida, Family Aulacopleuridae.
Description A trilobite with a compact exoskeleton and only slightly convex cephalon, much greater in width than in length. The glabella is only roughly half as long as the cephalon, and the result is a fairly long preglabellar area at the front of the cephalon. The glabella possesses two or three pairs of lateral furrows, the hindmost of which extends backwards as far as the occipital furrow posteriorly traversing the glabella. The eyes are small and well defined. The number of thoracic segments is highly variable, ranging from 12 to 22; the axial region of the thorax is narrow in comparison with the greater lateral development of the pleurae, which possess rounded tips and lack spines. The pygidium is of restricted length, with a variable number of segments; its margin is entire and carries no spines.
Stratigraphic position and geographical distribution The genus has a broad timespan: it first appears in the Middle Ordovician and lasts right through the whole Silurian up until the Middle Devonian (480–370 million years ago). It occurs in European and North African rock and in Greenland. The example in the photograph, some 2 cm (¾ in) long, comes from the Czechoslovakian Silurian at Lodenice.
Note The genus is subdivided into two subgenera which differ from each other in the position of the eyes in relation to the glabella and in the different proportions between the segments of the thorax and those of the pygidium.

127 CRYPTOLITHUS

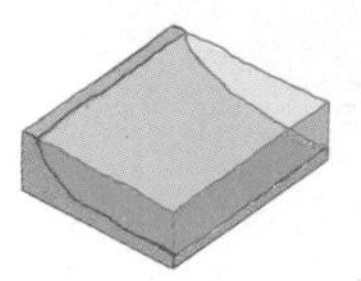

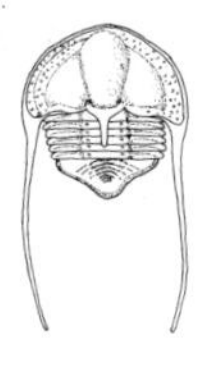

Classification Phylum Arthropoda, Class Trilobita, Order Ptychopariida, Family Trinucleidae.
Description An elegant form of trilobite with a very characteristic cephalon: it possesses an external region thickly covered in small, circular depressions, regularly arranged in a radial pattern both on the anterior part of the cephalon and on the sides. The genal spines are typically elongate: directed backwards, their length greatly exceeds the entire length of the body. The glabella and the cheeks are very swollen, with a globular glabella that extends as far as the anterior region. The thorax contains only six segments, with convex axial rings; the pleurae possess a diagonal furrow. The pygidium is subtriangular, centrally elongate, with rounded points; the axis of the pygidium is subdivided into several rings and the pleural area is slightly furrowed transversally.
Stratigraphic position and geographical distribution The genus ranges from the Lower to the Upper Ordovician (500–435 million years ago). It occurs in North America and England. The example in the photograph measures under a centimeter (½ in); it comes from the Ordovician of Pennsylvania and belongs to the species *C. geilulus*.
Note These trilobites have a curious way of enrolling: they fold their small thorax beneath their cephalon, bringing the posterior margin of the pygidium into contact with the anterior margin of the cephalon, thereby protecting most of their body beneath the head shield.

---

128 CALYMENE

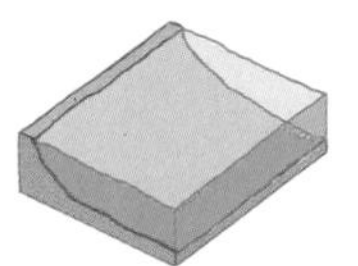

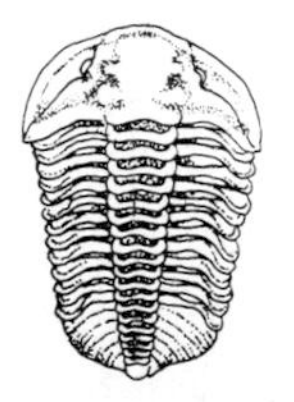

Classification Phylum Arthropoda, Class Trilobita, Order Phacopida, Family Calymenidae.
Description The cephalon of this trilobite is of semicircular form; it has a pronounced, convex glabella that is higher than the cheeks and folds downwards, extending anteriorly beyond them. The glabella also possesses three pairs of broad, deeply incised lobes. The thorax, which is well developed lengthwise, consists of 13 segments (only rarely 12); the axis, like the rest of the thorax, tapers posteriorly; the pleurae are rounded at the tip. The pygidium is short, with its axis subdivided into six clearly marked rings and six furrows in the pleural area; it is triangular in shape, with a rounded apex.
Stratigraphic position and geographical distribution The genus possesses a broad stratigraphic distribution, ranging from the Lower Silurian to the Middle Devonian (435–360 million years ago). It has a vast geographical range, occurring in Europe, Australia and North and South America. The example in the photograph belongs to the type species *C. blumenbachi*; it measures approximately 4 cm (½ in) and comes from the English Silurian.
Note This type of trilobite is able to enroll in the same way as many living arthropods. Representatives of this family were very abundant in Paleozoic seas from the Ordovician to the Devonian.

129 PHACOPS

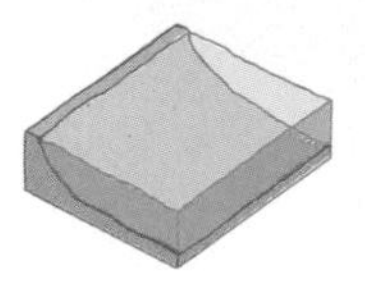

Classification Phylum Arthropoda, Class Trilobita, Order Phacopida, Family Phacopidae.
Description A typical form of trilobite, characterized by a posteriorly narrow glabella that widens conspicuously toward the front; the glabella has no pronounced furrows and in many forms is decorated overall by large granulations. The eyes are very large and often well preserved in fossils. It possessed uniramous antennae emerging near the hypostome, a plate present on the ventral region directly in front of the mouth of many trilobites. The thorax is well developed, with a broad axial region and round-tipped pleurae. The pygidium is semicircular and rounded, with three indentations on its posterior margin.
Stratigraphic position and geographical distribution The genus is widespread from the Silurian to the Devonian (435–345 million years ago), and this broad timespan is matched by a proportionally vast geographical distribution, occurring in Silurian and Devonian layers throughout the world. The example in the photograph, approximately 5 cm (2 in) long, belongs to the species *P. rana* and comes from the North American Devonian.
Note On the ventral side of its cephalon *Phacops* possesses a furrow, caused by a doublure, which served to house the posterior part of the pygidium when the animal was enrolled.

---

130 DALMANITES

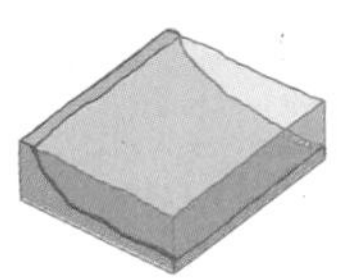

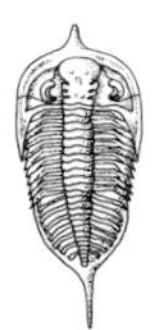

Classification Phylum Arthropoda, Class Trilobita, Order Phacopida, Family Dalmanitidae.
Description A very fine form of trilobite, with a compact exoskeleton possessing robust spines long and situated on the pygidium at its posterior and short at its anterior extremities. The cephalon is semicircular in shape, with a flattened and very distinct anterior margin. The glabella is highly developed and widens considerably toward the anterior; it is marked by three pairs of pronounced furrows dividing it into lobes in the posterior half. There is an elongate hypostome on the ventral region. The eyes are well developed and kidney-shaped. The genal spines, situated at the posterior extremities of the cephalon, are long, stout and retroflex. The thorax consists of ten to 12 segments, with pointed pleural spines. The pygidium is well developed in length and subdivided into 11 to 14 segments. There are six to seven pairs of ribs on the pleural area of the pygidium, which ends in a long spine.
Stratigraphic position and geographical distribution The genus is widespread from the Silurian to the Devonian (435–380 million years ago). It occurs in Europe, North and South America and Australia. The example in the photograph, some 5 cm (2 in) long, belongs to the type species *D. caudatus* and comes from the English Middle Silurian at Dudley.
Note The long spines with which *Dalmanites* is endowed have been interpreted as an aid to support the animal on the sandy sea bottom. This indicates that *Dalmanites* was a benthonic form.

131 ODONTOCHILE

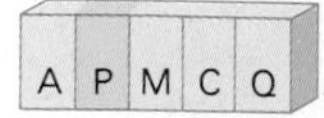

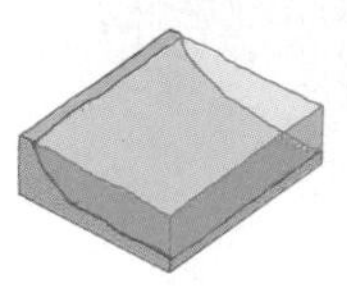

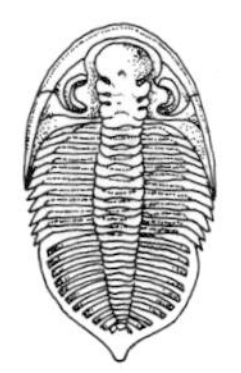

Classification Phylum Arthropoda, Class Trilobita, Order Phacopida, Family Dalmanitidae.
Description A trilobite bearing marked similarities to *Dalmanites*, though differing in several details. The cephalon is subelliptical in form, with two sturdy genal spines that fold sharply backwards and reach slightly beyond the middle of the thorax. The eyes are large and the glabella is elongate and broad in the anterior lobe, with four pairs of deep furrows that subdivide it into numerous lobes. There is a jagged hypostome on the undersurface of the cephalon. The thorax is subdivided into 11 segments, all of them of more or less the same width; the pleural spines are pointed at the tip. The pygidium, which is approximately as long as the thorax, is very well developed and possesses axial rings ranging in number from 16 to 22, as well as up to 15 pairs of ribs on the pleural region. It ends in a short spine, which is much less prominent than that of *Dalmanites*.
Stratigraphic position and geographical distribution The genus ranges from the Lower to the Middle Devonian (395–360 million years ago). It is widespread geographically, occurring in Europe, North and South America and Australia. The example in the photograph belongs to the type species *O. haussmanni* and comes from the Czechoslovakian Devonian at Locklow; it measures approximately 6 cm (2½ in).
Note The family Dalmanitidae contains numerous genera living from the Silurian to the Middle Devonian; they were very widespread, occurring in many different forms.

---

132 GREENOPS

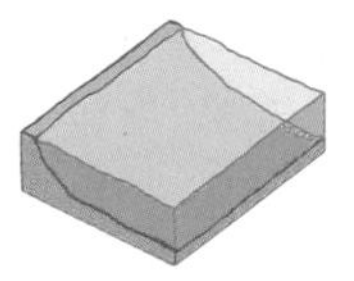

Classification Phylum Arthropoda, Class Trilobita, Order Phacopoda, Family Dalmanitidae.
Description This is a small trilobite with a slightly convex cephalon and straight anterolateral margins. The glabella is well developed, narrow at the back and broad at the front, and is marked by pairs of shallow furrows. The palpebral lobes are clearly visible at the sides of the glabella and the eyes are large. At the posterior end of the cephalon there are very pronounced, broad and retroflex genal spines. The pleurae on the thorax can sometimes develop long and spiny processes. The pygidium has a complex morphology, with an anteriorly broad axis tapering toward the posterior, and is formed of seven to ten rings, while the pleurae are markedly segmented, with five pairs of ribs and five pairs of broad, spiny appendages pointing either directly or obliquely backwards. At the tip of the pygidium there is a broad spiny appendage whose length does not exceed that of the preceding ones.
Stratigraphic position and geographical distribution The genus is typical of the Middle Devonian and has been doubtfully identified from the Upper Devonian (380–360 million years ago). It occurs in North America and Europe. The example in the photograph belongs to the species *G. collitelus* and comes from Erie County, New York.
Note The genus is divided into two subgenera on the basis of morphological differences in the thorax and pygidium: *Greenops* is typical of North America, while *Neometacanthus* is typical of central Europe.

133 CANADASPIS

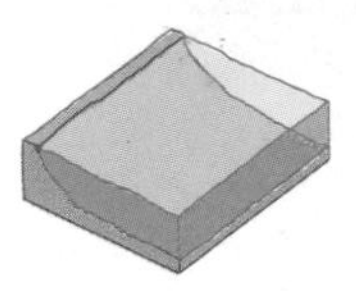

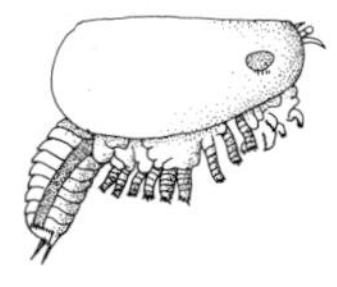

Classification Phylum Arthropoda, Superclass Crustacea, Class Malacostraca, Subclass Phyllocarida.
Description The anterior region of the body of this primitive crustacean has a well-developed carapace, which covers the cephalic region and the thorax, while the abdomen emerges posteriorly from it. The extremities of the limbs protrude from the lower part of the carapace. The carapace itself is composed of two lateral valves with an average length of approximately 35 mm (1½ in). The eyes and a pair of antennae are also visible. In fossils it is possible to distinguish the appendages that make up the oral apparatus as well. The thorax is divided into eight segments, each of which carries a pair of appendages, which in turn are composed of two elements. There are seven abdominal somites, all of them without appendages, apart from the last one. There is also a small telson.
Stratigraphic position and geographical distribution The genus contains a single species, *C. perfecta*, which is found exclusively in the Burgess Shale of British Columbia (Canada), dating back to the Middle Cambrian (500 million years ago). The carapace in the photograph measures 3.8 cm (1½ in).
Note In fossil form *Canadaspis* characteristically occurs in numerous groups. Because of the structure of its limbs it is probable that the animal lived on the sea floor, where it could move with ease, and possibly swim. Thanks to its powerful masticatory apparatus, it could eat firm pieces of food.

---

134 BALANUS

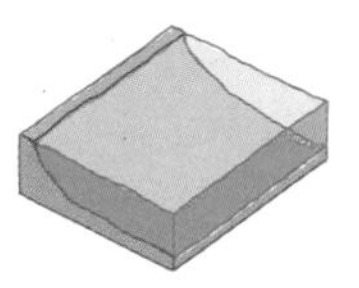

Classification Phylum Arthropoda, Superclass Crustacea, Class Cirripedia, Family Balanidae.
Description *Balanus* is a cirriped, a very specialized form of crustacean. The animal lives protected by a shell composed of four to six calcified plates, which may be fused together. These plates form a shell shaped more or less like a truncated cone, which is found in fossil-bearing layers either entire or split up into single plates. The shell has an orifice in its upper section to allow the animal to uncoil its appendages or "cirri," which are used for respiration and for catching food.
Stratigraphic position and geographical distribution The genus *Balanus* dates from the Eocene (55 million years ago) and is still found throughout the seas of the world in numerous subgenera and species. Its fossil distribution is equally broad and its remains are found almost everywhere in Tertiary and Quaternary sediments. The example in the photograph comes from the Italian Pliocene near Asti; it measures 4.5 cm (1¾ in).
Note Cirripeds attach themselves from the larval stage to a dead substratum such as rocks, pieces of wood and the keels of ships and boats. They live by filtering minuscule particles of food from the water; they are widespread in all seas, particularly in the zone lying between high and low tide.

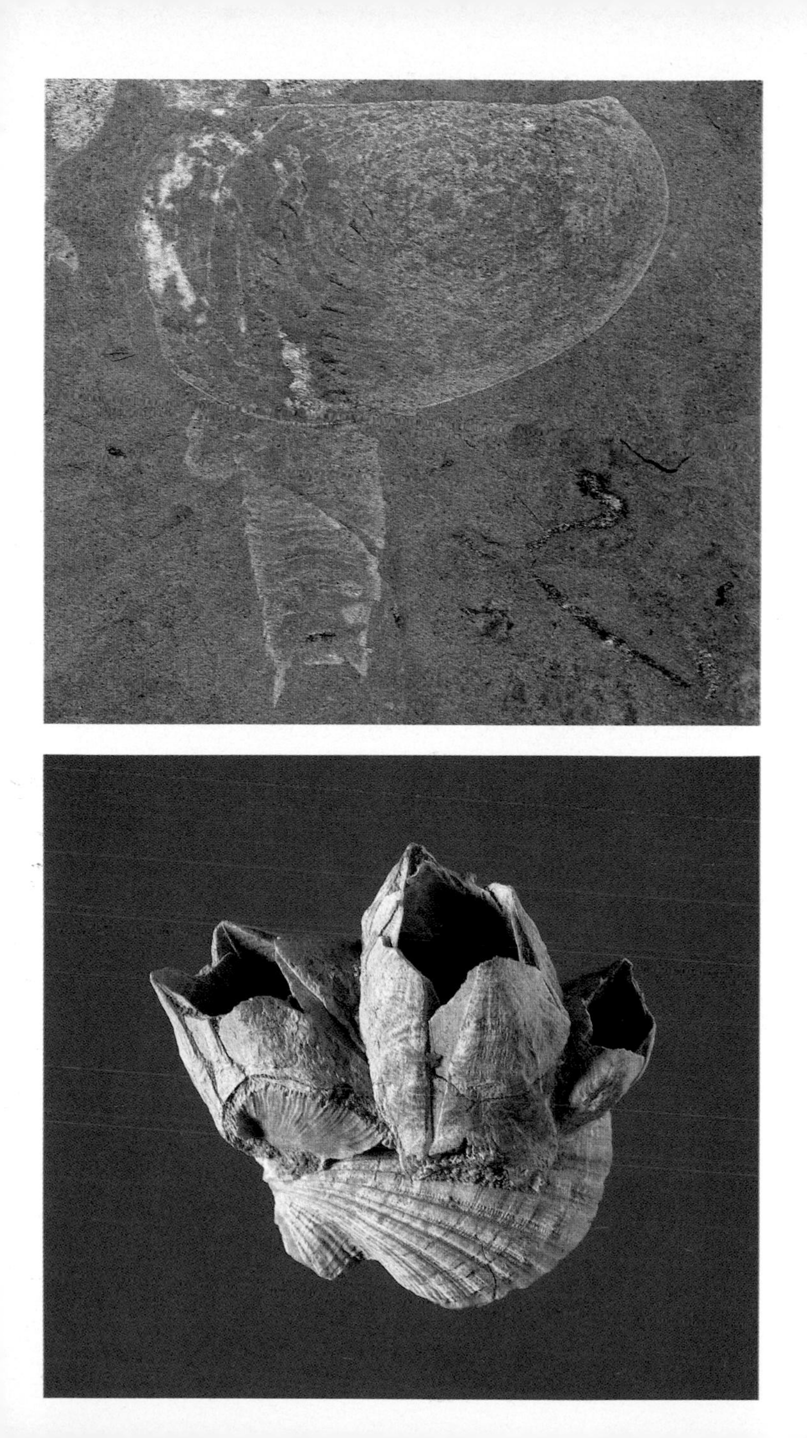

135 CERATIOCARIS

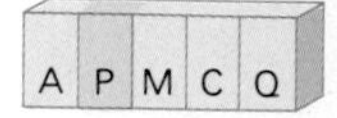

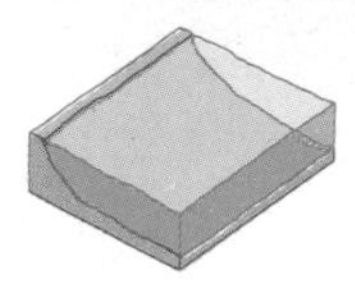

Classification Phylum Arthropoda, Superclass Crustacea, Class Malacostraca, Subclass Phyllocarida, Family Ceratiocaridae.

Description A phyllocarid crustacean with a carapace composed of two lateral valves, of the type found in other crustaceans retaining primitive characteristics, such as Branchiopoda and Copepoda; unlike the latter, however, the phyllocarid's whole body is not contained within the carapace, only the anterior section. In addition, phyllocarid carapaces have an anterior mobile plate or a rostral plate, one of their unique characteristics. In *Ceratiocaris* the carapace is in the form of an elongate oval, with a very sturdy anterior rostrum, covered, like the rostral plate, in dense longitudinal striae. The abdominal somites are anteriorly covered by tilelike scales. The telson has two elongate appendages and a caudal furca covered in spines. The limbs, where present, are poorly preserved in fossil form; this makes reconstruction impossible.

Stratigraphic position and geographical distribution The genus ranges from the Upper Ordovician to the Upper Devonian (450–370 million years ago). Examples of *Ceratiscaris* occur in Paleozoic strata worldwide. The example is from the Scottish Devonian at Lesmahagow.

Note Modern phyllocarids are usually small in size, achieving a maximum length of approximately 4 cm (1½ in). Fossil examples, by contrast, can be much larger: the one in the photograph is approximately 15 cm (6 in) long; other examples have been discovered measuring 75 cm (30 in) in length.

---

136 PSEUDARCTOLEPIS

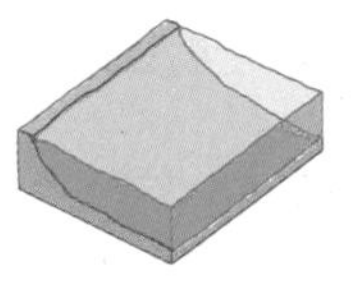

Classification Phylum Arthropoda, Superclass Crustacea, Class Malacostraca, Subclass Phyllocarida.

Description Only the carapace is known of this phyllocarid, as is often the case with fossil Phyllocaridae. As a consequence serious difficulties are encountered in interpreting its remains, which may sometimes become confused with those of other organisms or prove impossible to fit satisfactorily into the group's systematics. The carapace of *Pseudarctolepis* is of elongate form, with a beaklike elongation on the dorsal region, both anteriorly and posteriorly. Another characteristic feature of this genus is the long spine located at the front of the ventral region and curving backwards to extend below the carapace.

Stratigraphic position and geographical distribution This is a very ancient form of phyllocarid, dating as far back as the Middle Cambrian (530 million years ago), and is typical of the rocks of the Wheeler Formation in the House Range, Utah.

Note Modern phyllocarids are benthonic organisms that live in the mud or beneath pebbles in shallow waters, although there are species living at depths of more than 2,000 meters (6,560 ft). The capacity for active swimming is known in only one living species. It is believed that the majority of fossil phyllocarids were also benthonic forms, which lived on the sea floor, filtering minuscule particles of food. Perhaps some forms were capable of active swimming or were at least planktonic.

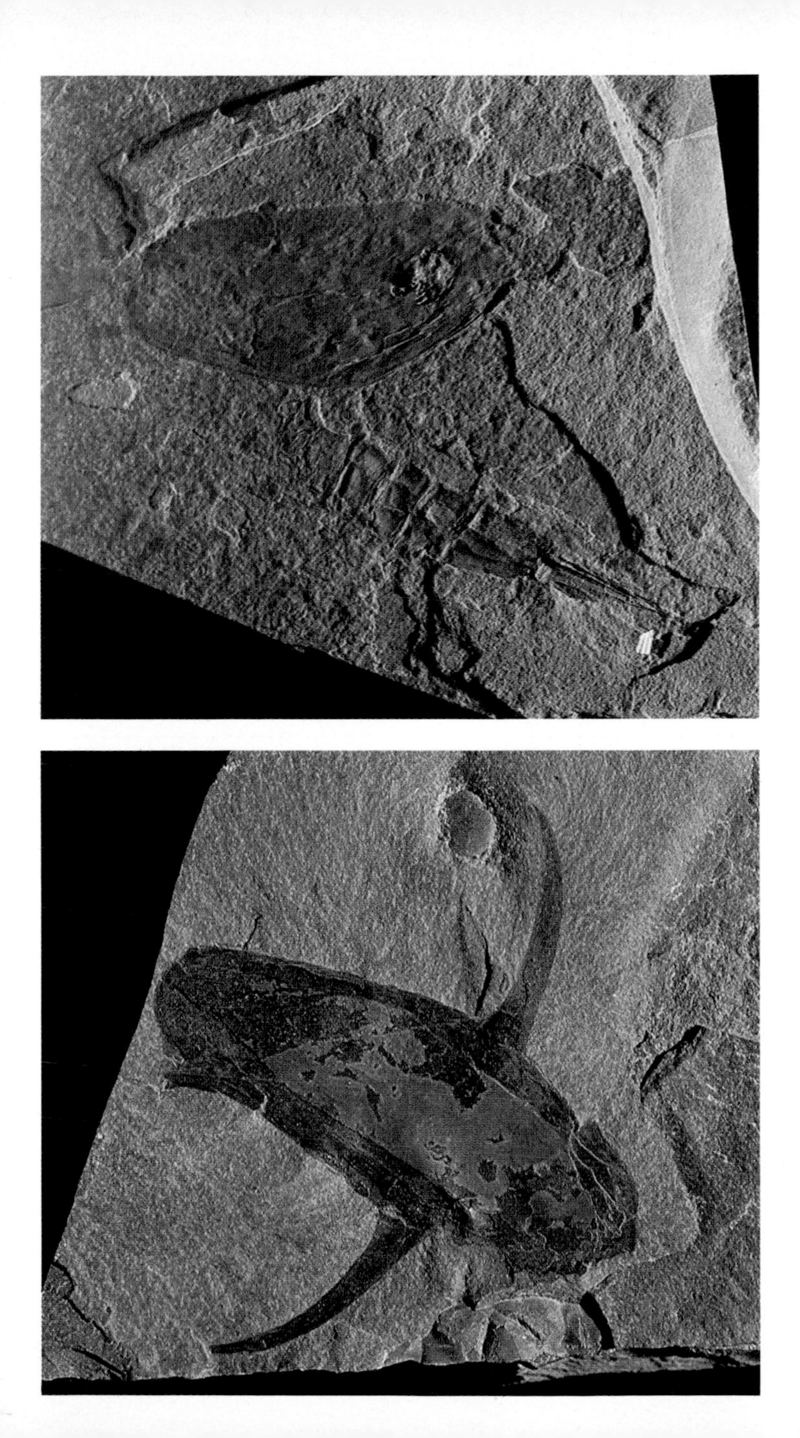

137 ATROPICARIS

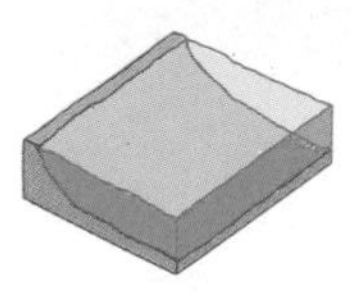

Classification Phylum Arthropoda, Superclass Crustacea, Class Thylacocephala.
Description An unusual form of crustacean, the most obvious characteristic of which is its thin, univalve carapace, up to 25 mm (1 in) long and trapezoidally-shaped when viewed from the side, with a dorsal rostrum on its anterior section. The carapace is decorated for its entire length by a continuous series of small, vertical and falciform laminae. Like all crustaceans, the members of Thylacocephala so far identified have no eyes; the head is modified to form a sort of closed bag emerging anteriorly from the carapace. The rest of the body is contained within the carapace, from which emerge only three pairs of segmented cephalic appendages, the thoracic appendages, shaped like small shovels and part of the abdomen, which is very short and rudimentary. There are highly developed gills inside the carapace.
Stratigraphic position and geographical distribution The genus contains a single species, *A. rostrata* (see photograph), which comes from the Rhaetian (Upper Trias, 200 million years ago) of the Alpine foothills in Lombardy (Italy).
Note The class Thylacocephala, discovered only recently, comprises strange crustaceans forms, exclusively fossils, some of whose species can reach over 30 cm (12 in). They lived in muddy waters, could hardly move, and fed, in some cases at least, on the dead bodies of other organisms—fish, sharks or fellow crustaceans. Their remains are known from the Devonian to the Cretaceous.

---

138 BELOTELSON

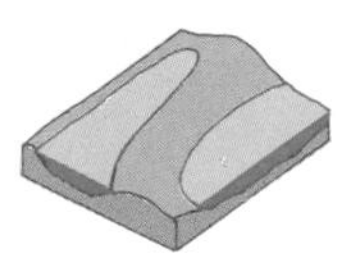

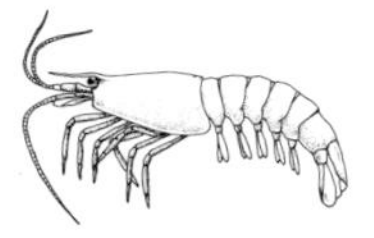

Classification Phylum Arthropoda, Superclass Crustacea, Class Malacostraca, Order Eocarida, Family Belotelsonidae.
Description An eocarid crustacean of modest dimensions, with a light, elongate carapace from which there emerges, anteriorly, a rostrum whose length can be equivalent to as much as a third to a half of the carapace's total length. *Belotelson* possesses antennules consisting of a segmented pedicle supporting two short flagella, and antennae, equipped with a single, elongated flagellum. The small eyes are supported on a pedicle. Below the carapace there are well-developed thoracic legs which, unlike those of decapod crustaceans, number 16 instead of ten. The abdomen of this crustacean is broad and as long as the carapace; the abdominal appendages are biramous and paddle-shaped. The abdomen also ends in a narrow, pointed triangular telson, with fanlike uropoda at the sides; at the end of its telson there are two small caudal lobes.
Stratigraphic position and geographical distribution The genus is known in the Carboniferous (345–280 million years ago). It occurs in the English Coal Measures (Upper Carboniferous), but is most abundant in the Upper Carboniferous of Mazon Creek, Illinois; it is also found in Montana and Indiana. The example, type species *B. magister* from Mazon Creek, measures approximately 3 cm (¼ in).
Note From its general morphological characteristics, *Belotelson* must have been a benthonic organism, moving over the sea floor by its thoracic appendages.

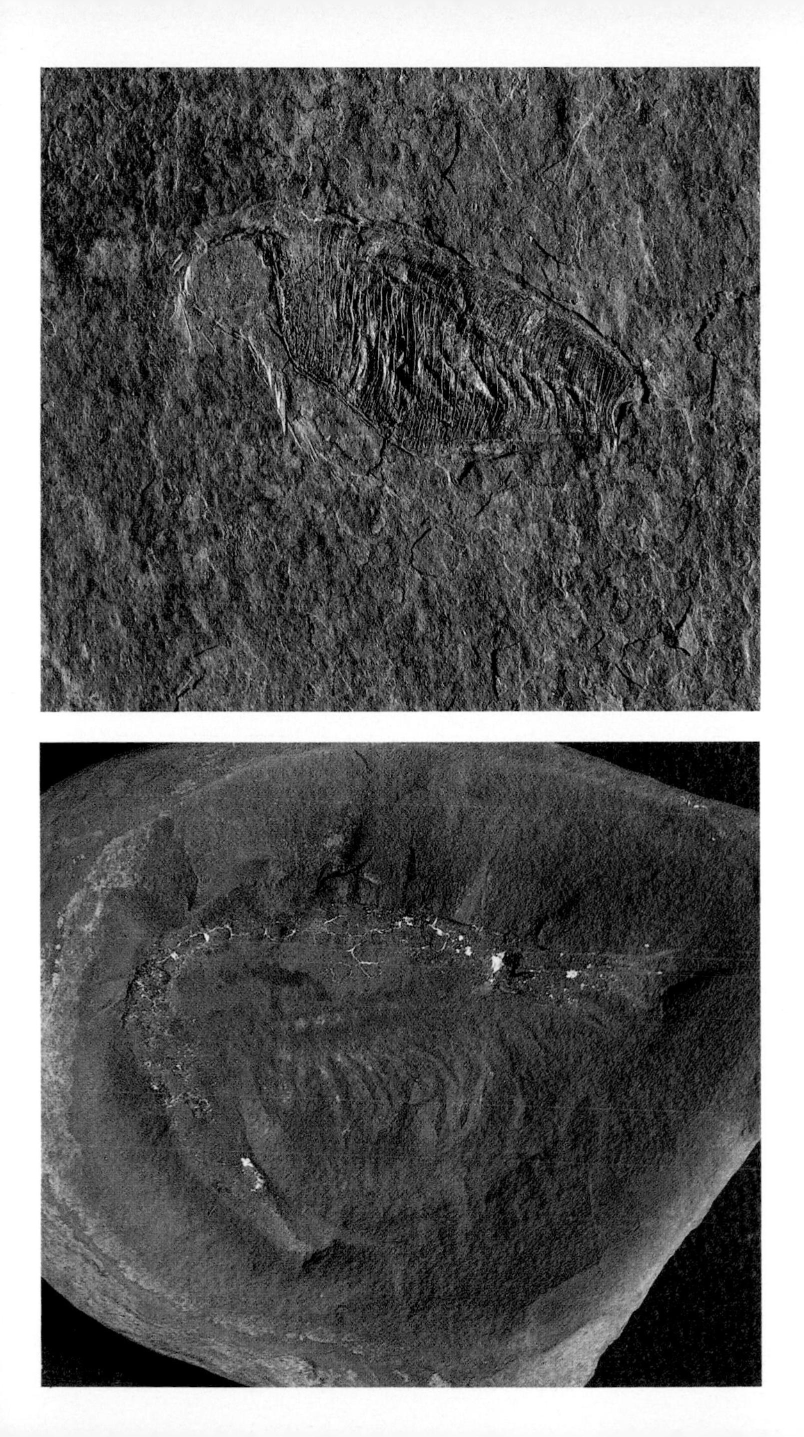

139 ANTRIMPOS

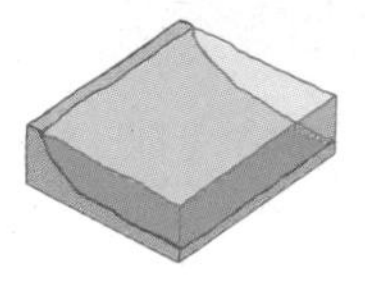

Classification Phylum Arthropoda, Superclass Crustacea, Class Malacostraca, Order Decapoda, Family Penaeidae.
Description This decapod crustacean is like modern *Penaeus*. The external skeleton is composed of an elongate, univalve carapace covering the anterior part of the body; this had a spine or antennal spine on its anterior section. The antennules are very short, the antennae very elongate, up to twice the the body length. The thoracic appendages are very long; the first three pairs of legs, covered in spines, have chelae at the tip. The abdomen is well developed and subdivided into segments, the last of which is also the longest. The triangular, pointed telson is crossed by a longitudinal furrow. The uropodite of the outermost appendage (the exopodite) has a dieresis.
Stratigraphic position and geographical distribution The genus has a wide stratigraphic distribution from the Permian to the Cretaceous (250–70 million years ago); its geographical distribution is limited to Europe and Madagascar. In Europe it sometimes abounds in the Triassic of the Alpine foothills in Lombardy (Italy) and in the Upper Jurassic of Solnhofen, Bavaria. The photograph shows *A. noricus*, about 5 cm (2 in) long, from the Valvestino near Brescia (Italy).
Note Penaeid crustaceans are very widespread in modern seas; as fossils their delicate remains (the exoskeleton, for example, has several component elements) are rare, although in some strata many carapaces are found.

---

140 AEGER

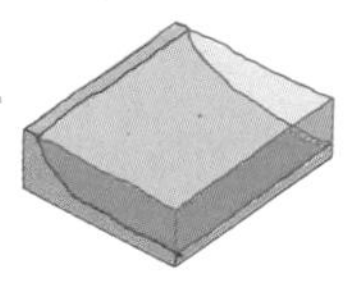

Classification Phylum Arthropoda, Superclass Crustacea, Class Malacostraca, Order Decapoda, Family Penaeidae.
Description A crustacean typical of Jurassic marine sediments, where it is found in numerous species. Its form is typical of swimming decapods, with an elongate, univalve carapace covering the anterior portion of the body, while posteriorly there emerges the long abdomen, possessing six somites and a pointed, triangular telson. The cephalic and thoracic appendages are highly developed: the biramous antennules are long and the two very long antennae, despite their delicacy, may be perfectly preserved in certain exceptional strata. The third maxilliped is highly developed and can exceed the dimensions of the other thoracic appendages, which act as swimming legs. This third maxilliped is equipped with spines (as are some of the legs), sometimes of considerable size. The first three pairs of legs are equipped with chelae at their extremities.
Stratigraphic position and geographical distribution Species of the genus range from the Upper Triassic to the Upper Jurassic (200–140 million years ago). Despite the delicacy of the exoskeleton, examples very well preserved in every detail have been found in strata such as those at Osteno (Italy) and Solnhofen (Germany). The example in the photograph is from the Sinemurian at Osteno (the lower part of Lake Lugano in Lombardy) and measures approximately 15 cm (6 in).
Note *Aeger* was a swimming, marine decapod inhabiting warm, shallow waters.

141 ERYMA

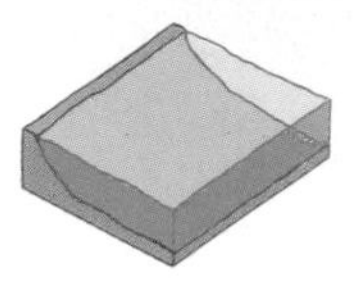

Classification Phylum Arthropoda, Superclass Crustacea, Class Malacostraca, Order Decapoda, Family Erymidae.
Description A decapod crustacean, very similar to the modern lobster. The carapace is strong, oval when viewed from the side and cylindrical overall, with a small anterior rostrum. It is covered with dense tubercles that extend over the first pair of legs and the abdomen. The first pair of legs (pereiopods) is more developed than the other four pairs: it is very robust and possesses strong, stumpy chelae. The second and third pairs of pereiopods are also equipped with chelae, but these are much smaller. The animal possesses antennules, short and biramous, and antennae, which are more elongate. The telson is triangular, with rounded sides, and the exopodite on the uropod (the uropods are fanlike expansions beside the telson at the end of the body) possesses a suture or dieresis.
Stratigraphic position and geographical distribution The genus ranges from the Lower Jurassic to the Upper Cretaceous (190–70 million years ago) and has a broad geographical distribution, occurring in certain locations in Germany, Italy and France and in East Africa, Indonesia and North America. The example in the photograph, approximately 6 cm (2½ in) long, belongs to the type species *E. modestiformis*, characteristic of the Solnhofen beds in Bavaria, dating from the Upper Jurassic.
Note *Eryma* was a marine crustacean that lived in shallow depths.

---

142 HOPLOPARIA

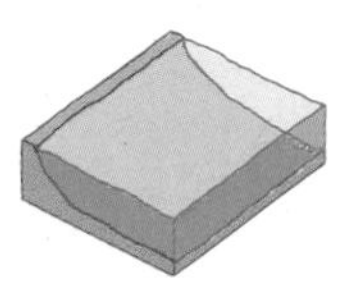

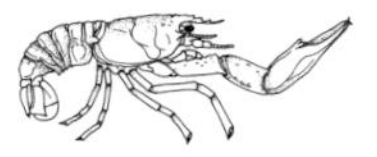

Classification Phylum Arthropoda, Superclass Crustacea, Class Malacostraca, Order Decapoda, Family Nephropsidae.
Description Crayfishlike decapod characterized by a primary pair of very developed pereiopods equipped with stout chelae. The carapace is cylindrical, with clearly visible furrows: these furrows are of great systematic significance in fossil decapods; their development and their shape represent an important characteristic for classification at the genus level and make it possible to identify even when the fossil consists solely of the carapace and lacks a complete exoskeleton, as is frequently the case with crustaceans. The antennae are well developed; the anterior rostrum of the carapace is thin and may be serrated; the ornamentation is made up of tubercles. There is a remarkable resemblance between *Hoploparia* and the living *Homarus*: so much so, in fact, that it is sometimes hard to attribute fossil species to one genus or the other.
Stratigraphic position and geographical distribution The genus ranges from the Lower Cretaceous to the Lower Tertiary (140–50 million years ago) and is distributed throughout the world as far as the Antarctic. The example in the photograph, approximately 12 cm (5 in) long, comes from the Lower Cretaceous at Lyme Regis, Dorset (England).
Note Remains of *Hoploparia* can sometimes be preserved at the moment of molting, when the animal abandons its old skeleton to form another, larger one. After each molt the animal's body size increases by about a quarter.

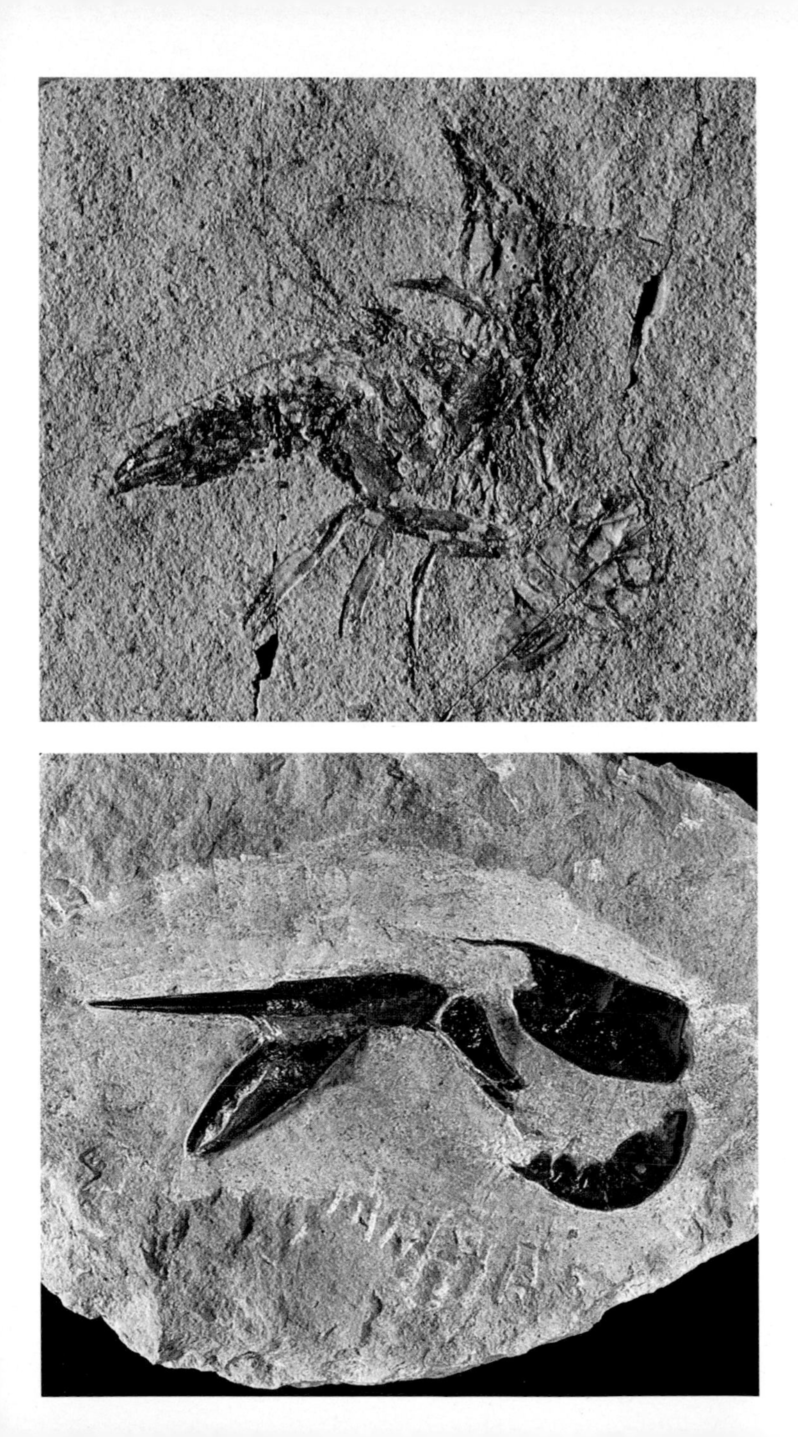

143 HOMARUS

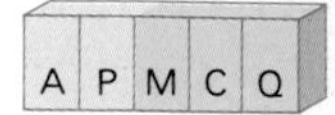

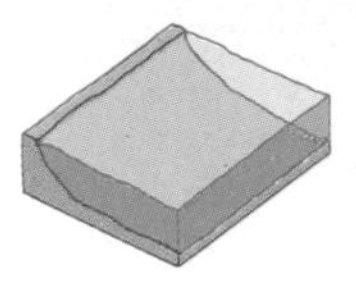

Classification Phylum Arthropoda, Superclass Crustacea, Class Malacostraca, Order Decapoda, Family Nephropsidae.
Description A living crustacean, although the description given here applies to its known fossil forms, dating back to the end of the Mesozoic. The carapace is short and deep, with a small, serrated rostrum. The antennules are biramous and short, whereas the antennae are much more elongate and composed of a single flagellum. The first pair of thoracic legs bear very long, sturdy chelae at their extremities, very similar to those of the living *H. gammarus*. There are certain similarities between the fossil forms of *Homarus* and *Hoploparia*, but this genus differs from *Hoploparia* in the dimensions of the mobile section of its chelae (the dactylus) and the smaller overall size of its body.
Stratigraphic position and geographical distribution The genus ranges from the Cretaceous (140 million years ago) to the present; it is now widespread in European and North American seas, while in fossil form it occurs in the Cretaceous in North America, Europe and the Lebanon. The example, *H. hakelensis*, is from the Lebanese Cretaceous at Hakel; it measures about 8 cm (3¼ in).
Note The fossil species of the genus are small in size, whereas the modern *H. gammarus* can reach a length of some 56 cm (22½ in). The lifestyle of the fossil forms must have been analogous to those of the living forms, favoring shallow, rocky waters. They also burrow in sandy sediments. *H. gammarus* is highly sought after for its delicious meat.

---

144 COLEIA

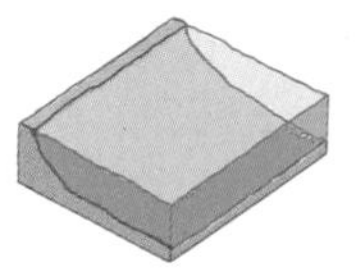

Classification Phylum Arthropoda, Superclass Crustacea, Class Malacostraca, Order Decapoda, Family Coleiidae.
Description A decapod crustacean of the superfamily Eryonidae which, unlike the majority of other related decapods, is of compressed, broad and flattened shape. The carapace is subtrapezoidal, with longitudinal keels and two marked furrows traversing its entire width; the carapace may be ornamented with locally diffuse granulation and spines; the frontal margin may possess pronounced supraantennal spines. Both the antennules and the antennae are short. The telson is triangular and the exopodite on the uropods bears a characteristic dieresis.
Stratigraphic position and geographical distribution The genus is known throughout the Jurassic up until the Lower Cretaceous (190–100 million years ago). It is found in Europe (England, France, Germany, Italy), in western Siberia and in India. The example in the photograph belongs to the species *C. mediterranea*, measures approximately 12 cm (5 in) and comes from the Italian Lower Jurassic at Osteno, on Lake Lugano (Switzerland).
Note The decapod Eryonidae, of which *Coleia* and *Eryon* are two of the most typical representatives, was very abundant in the shallows of Mesozoic seas. A very few forms still survive living in deep waters at a depth of 200 meters (656 ft) and only occasionally climbing back onto the continental shelf. These decapods possess unique larval forms, known as *Eryoneicus*, which reach lengths of up to 7 cm (2¾ in).

145 PHALANGITES

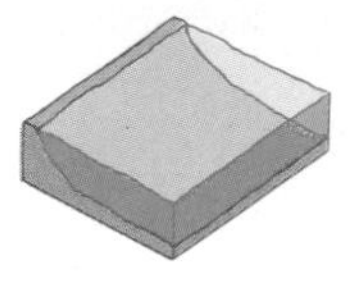

Classification Phylum Arthropoda, Superclass Crustacea, Class Malacostraca, Order Decapoda, Family Palinuridae.
Description This is a problematical organism, characterized by thin, elongate limbs like those of certain spiders. Its systematic classification has been the subject of considerable debate in the past: it is now believed that *Phalangites* represents the larval form of a decapod crustacean, probably *Palinurina*, a small primitive marine crayfish discovered in the same layers as the remains of *Phalangites*.
Stratigraphic position and geographical distribution *Phalangites*, in the form of the species *P. priscus*, dates from the Kimmeridgian (Upper Jurassic, 140 million years ago) at Solnhofen, Bavaria. The example in the photograph is a member of this species; it measures 3 cm (1¼ in).
Note *Phalangites* represents the only known fossil larva of a decapod, even though its attribution remains doubtful. Decapods, once their eggs have hatched, pass through various larval stages, each of which has its own name. In the family Palinuridae the first postembryonic larval stage is known as a phyllosome and has a discoidal body with long, thin legs: it is at this stage that *Phalangites* has been identified.

---

146 HARPACTOCARCINUS

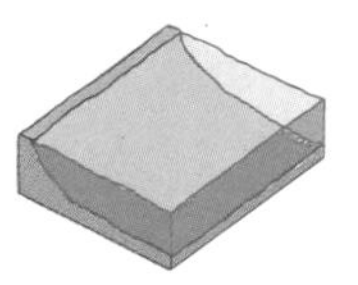

Classification Phylum Arthropoda, Superclass Crustacea, Class Malacostraca, Order Decapoda, Family Xanthidae.
Description A crab with a carapace of rounded outline, much greater in width than length. The back is moderately convex, punctuate and not markedly divided. The rounded anterolateral margin and the front are both equipped with spines. The chelae are strong and of different sizes in the same animal (heterochelate), with spiny tubercles in the dorsal region of the final section (propodus).
Stratigraphic position and geographical distribution The genus ranges from the Lower Eocene to the Upper Eocene (55–35 million years ago). It has a fairly large geographical spread, occurring in Europe, East Africa and the state of Texas. Extremely fine fossils, often in their entirety, are found in Eocene sediments of the Veneto region in Italy. One such example can be seen in the photograph, a member of the type species *H. punctulatus*; the width of this specimen's carapace is approximately 8 cm (3¼ in).
Note Crabs, or decapod Brachyura, are first known from the Jurassic. The hypothesis put forward by paleontologists on the basis of fossil evidence is that they are derived from the glyphaeid decapods Palurinidae, which were widespread during the Mesozoic and are now represented by a single species found in the Philippines.

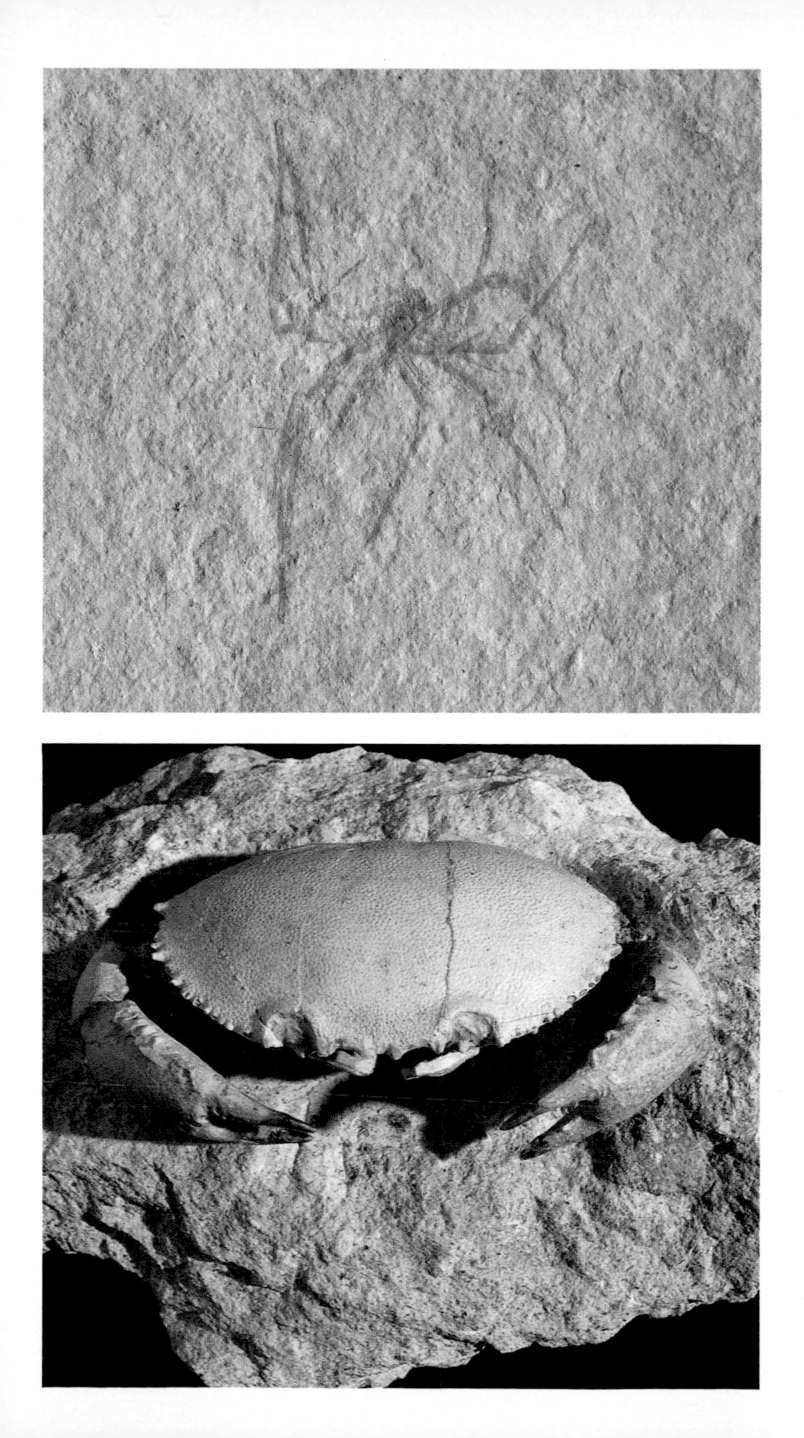

147 ZANTHOPSIS

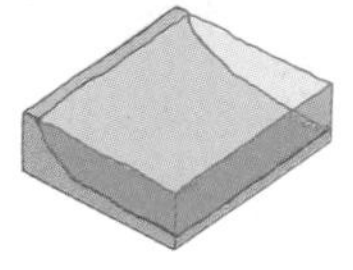

Classification Phylum Arthropoda, Superclass Crustacea, Class Malacostraca, Order Decapoda, Family Xanthidae.
Description Crabs of moderate size, with an oval, dorsally convex carapace; the front possesses prominent spines around the orbital region. The sides, on which there are teeth, are sharp; the chelae are short and sturdy, with their sizes differing in individual specimens, a phenomenon known as heterochelism.
Stratigraphic position and geographical distribution The genus is known from the Paleocene to the Oligocene (65–22.5 million years ago). Its distribution in rocks of the Lower Tertiary is broad: it is found in Europe, Africa and North and Central America. The example in the photograph belongs to the species *Z. vulgaris* and comes from the Oligocene in the state of Washington; it measures approximately 12 cm (5 in).
Note Crabs are often found in rounded nodules in clay: there is fine three-dimensional preservation of the exoskeleton in the specimen in the photograph, which survives perfectly in every detail.

---

148 ARCHAEOCYPODA

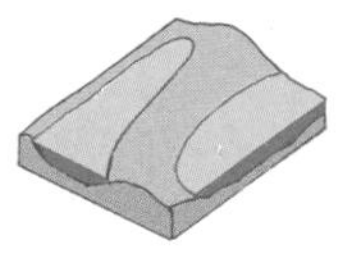

Classification Phylum Arthropoda, Superclass Crustacea, Class Malacostraca, Order Decapoda, Family Ocypodidae.
Description A medium-sized crab, with a carapace slightly wider than it is long, which can measure up to approximately 6 cm (2½ in) in length. The front is narrow and the eyes are borne on pedicles. Between its rounded margins and also dorsally, the carapace displays a fairly marked subdivision into different regions. The first pair of legs is equipped with a pair of long, robust chelae. These chelae are crossed by a longitudinal keel characteristically marked by finely tuberculate, transversal striae.
Stratigraphic position and geographical distribution The genus *Archaeocypoda*, and especially the species *A. veronensis* (see photograph), is known in the famous Monte Bolca beds found north of Verona (Italy), dating from the Middle Eocene (45 million years ago).
Note This genus is an ancient representative of the Ocypodidae. In the arrangement of the appendages, in its morphology and in certain characteristics of the abdomen, it displays notable affinities with such living representatives of the family as *Ocypoda*. It probably lived near the shore, onto which it climbed in search of food at low tide: the strong chelae acted both as a means of tearing off strips of food and also of carrying them to its mouth. At high tide it would take refuge beneath pebbles or burrowed in the sand.

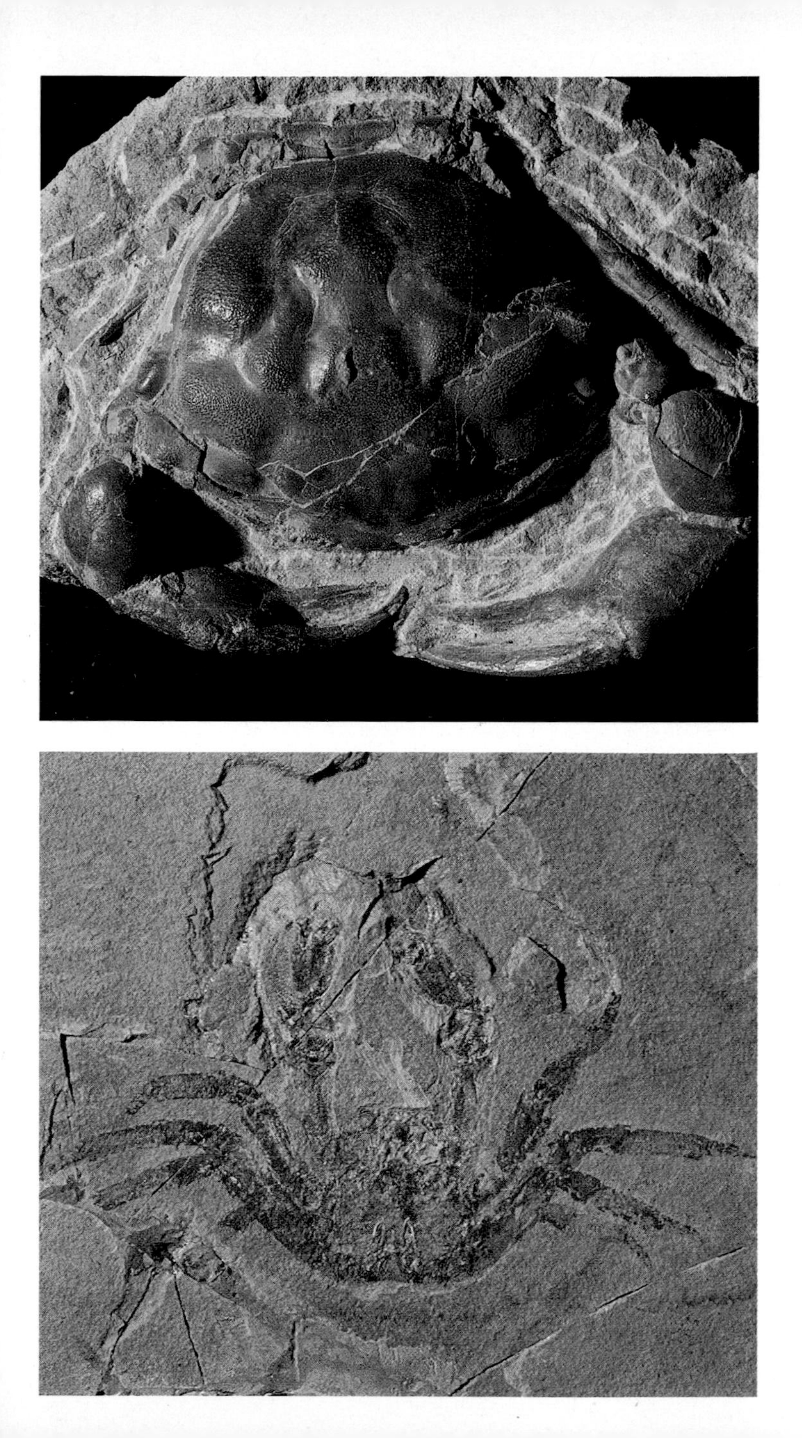

149 KALLIDECTHES

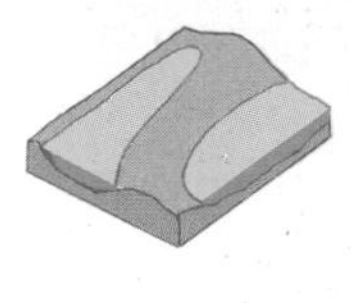

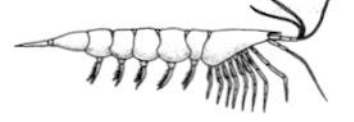

Classification Phylum Arthropoda, Superclass Crustacea, Class Malacostraca, Superorder Hoplocarida, Order Palaeostomatopoda, Family Kallidecthidae.
Description A primitive form of hoplocarid crustacean, with remote affinities with the modern *Squilla*. The animals did not exceed 7 cm (2¾ in) from the tip of the carapace to the posterior end of the uropods. The carapace is of subtriangular outline, with a smooth surface and a small, mobile rostrum on the anterior region in a dorsal position. The thoracic appendages, below the carapace, consist of eight pairs of legs decreasing in length toward the posterior region. The segmented pedicle of the first antenna bears three short flagella of different lengths, whereas the second antenna bears a single flagellum. The abdomen consists of six segments, the first five with biramous appendages. The telson is shovel-shaped, with a very wide posterior margin; beside it are biramous uropods, whose narrow outermost appendage (exopodite) extends backwards beyond the telson.
Stratigraphic position and geographical distribution The genus occurs in the Pennsylvanian stratum (Upper Carboniferous, 300 million years ago) in the state of Illinois. The example in the photograph, *K. richardsoni*, is from Mazon Creek, Illinois, and measures about 5 cm (2 in).
Note The thoracic appendages of this crustacean were adapted for swimming; there is no development of the first pair of legs into toothed appendages for seizing prey, typical of hoplocarids from the Mesozoic on.

---

150 PSEUDOSCULDA

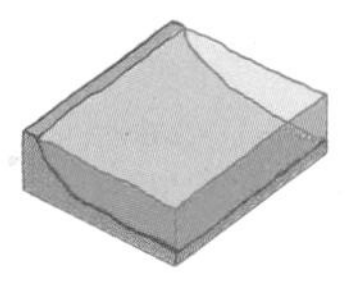

Classification Phylum Arthropoda, Superclass Crustacea, Class Malacostraca, Order Stomatopoda, Family Sculdidae.
Description A stomatopod crustacean of small size (the maximum is some 55 mm or 2¼ in) which resembles the modern *S. mantis* or squill. The carapace is short and only partially covers the thorax, which emerges posteriorly in front of the abdomen. One characteristic feature, as in the other stomatopods, is the development of the second pair of thoracic appendages, which take the form of pincers and are equipped with long spines on the inner part of the last segment. The antennules and the antennae are short. There are six thoracic segments (somites) and each bears a spine on the center of the back. The broad telson is trapezoidal and serrated, with two mobile spines at its apex. The exopodite of the uropodite bears a row of mobile teeth on its external margin and the uropods possess a bifurcated appendage.
Stratigraphic position and geographical distribution The genus is known in the Lebanese Cenomanian (Upper Cretaceous, 100 million years ago). The example in the photograph comes from the Cenomanian at Hakel (Lebanon) and belongs to the species *P. laevis*.
Note Like the modern *S. mantis*, which is caught by fishermen for its delicious meat, the carnivorous *Pseudosculda* was an inhabitant of muddy shallow bottoms where it dug the tunnels in which to live.

151 EURYPTERUS

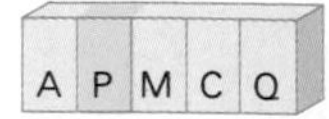

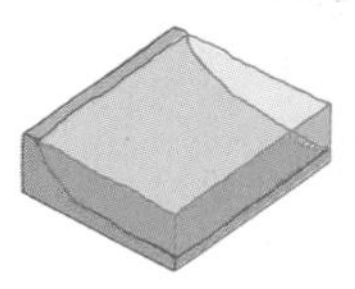

Classification Phylum Arthropoda, Class Merostomata, Subclass Eurypterida, Family Eurypteridae.
Description A typical genus of Paleozoic eurypterid arthropod whose body consists of an anterior region (prosoma) of subtrapezoidal shape and an elongate posterior region. The prosoma has six pairs of appendages: the first pair form the antennae, the next four pairs are masticatory appendages and the last pair, paddle-shaped, are for swimming. Compound eyes occupy a dorsal position. The posterior part of the body consists of an abdomen, or opisthosoma, divided into twelve segments, the first of which are wide then gradually tapering toward the rear; at the posterior extremity an elongate telson forms a robust spine.
Stratigraphic position and geographical distribution The genus ranges from the Ordovician to the Lower Permian (440–260 million years ago). The example in the photograph, some 10 cm (4 in) long, belongs to the type species *E. remipes* and comes from the Upper Silurian in the state of New York.
Note Eurypteridae were marine or brackish water organisms that enjoyed a rapid development from the Upper Silurian to the Lower Devonian. They include the largest known arthropod forms: specimens have been discovered measuring as much as 2 meters (6.5 ft) long, and in some cases they may even exceed 2.5 meters (over 8 ft). They were mainly benthonic, but there were also several free-swimming forms. They were all carnivorous predators, feeding off any type of prey they could exploit, including primitive ostracoderms.

---

152 MESOLIMULUS

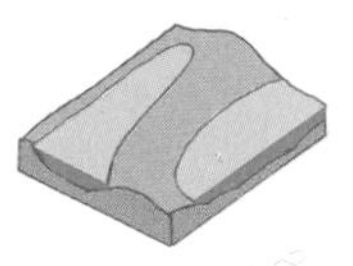

Classification Phylum Arthropoda, Class Merostomata, Subclass Xiphosura, Family Mesolimulidae.
Description *Mesolimulus* possesses an exoskeleton formed of a wide and very convex carapace, in the shape of a terminally elongate semicircle; it consists of a central glabella situated between two cheeks which extend backwards to form stout points; the compound eyes are situated laterally on the cheeks. The mouth, ventrally placed, typically bears a pair of appendages, or chelicerae, and five pairs of masticatory legs; the abdomen bears five pairs of legs and ends in an extremely long telson with a robust spine.
Stratigraphic position and geographical distribution The genus is known in the Upper Jurassic (Kimmeridgian stage, 140 million years ago) at Solnhofen, Bavaria, from where the example in the photograph originates. There are traces of the animal's movement on the surface of the rock: these take the form of a spiral circle, with the remains of the animal situated at the end. This phenomenon has been attributed to the organism having been asphyxiated in a very poorly oxygenated environment, losing its sense of direction and turning in a circle before dying. It measures approximately 12 cm (5 in).
Note *Mesolimulus* displays marked similarities with the living *Limulus*, the sole xiphosurid arthropod to have survived up until the present and which has come to be regarded as a real "living fossil." It lives in littoral waters off the Atlantic coast of North America and off the Pacific coast of Asia; it is a predominantly carnivorous animal.

153 CYCLUS

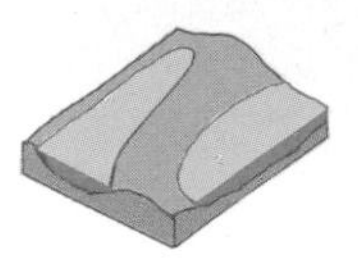

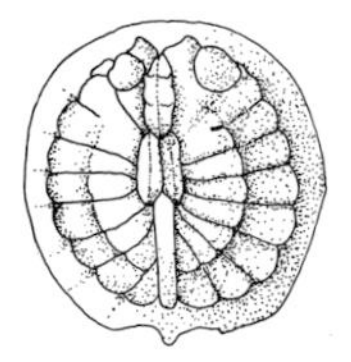

Classification Phylum Arthropoda, Family Cyclidae.
Description This is a strange small organism whose entire body is covered by a flattened carapace, either circular or oval and crossed in the middle of its back by a small longitudinal keel. The surface of the carapace may be either smooth or finely granulated. On the frontal margin two antennae emerge in a lateral direction. The ventral margin of the carapace reveals the remains of numerous pairs of legs (perhaps seven), which were in all probability biramous. The tiny abdomen is represented by a small telson and a pair of lobate appendages that are also small in size.
Stratigraphic position and geographical distribution *Cyclus* ranges from the Lower Carboniferous to the Lower Permian (345–270 million years ago). It has a wide geographical distribution, occurring in Carboniferous strata in northern Europe (Belgium, England, Ireland), in the Urals, Central Asia, and the states of Illinois and Kansas; it is also found in the Permian in the Urals and Sicily. The example in the photograph, approximately 1 cm (½ in) long, belongs to the species *C. americanus* and comes from the Illinois Upper Carboniferous.
Note *Cyclus* and its related genera have come to be regarded as crustaceans, even though they cannot be linked to any known group; they nevertheless also display affinities with trilobites and xiphosurids, such as the living *Limulus*. The flattened shape of their shells indicates that they must have been benthonic organisms living in shallow sea waters.

---

154 LIBELLULIUM

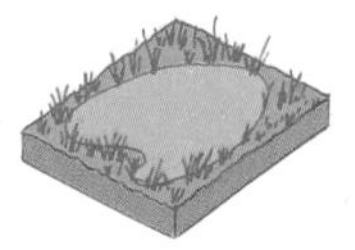

Classification Phylum Arthropoda, Class Insecta, Order Odonata.
Description This is a well-known type of preservation of a fossil dragonfly. Dragonflies are insects equipped with wings, with a body subdivided into three parts—head, thorax, abdomen—with the mandibles, the antennae and two pairs of jaws situated on the head. The thorax is composed of three fused segments, each of which bears a pair of legs; the second and third segments each bear a pair of wings. These wings are membranous and supported by a chitinous venation of characteristic design. The posterior part of the body, or abdomen, consists of 11 segments, some of which are modified to fulfill a sexual function, and is elongate in appearance.
Stratigraphic position and geographical distribution *Libellulium* is known through specimens that have been miraculously well preserved, despite their original delicacy, in the Kimmeridgian (Upper Jurassic, 140 million years ago) at Solnhofen, Bavaria. Forms of the genus *Cymathophlebia* have also been attributed to the genus *Libellulium* by synonymy. The example in the photograph, which measures approximately 5 cm (2 in), belongs to the species *L. longialata*.
Note Modern dragonflies are insects that live close to lakes or ponds; good fliers, they are carnivorous organisms that feed on tiny creatures. Dragonflies are known in fossil form from the Carboniferous, a period during which there was one typical form of gigantic size, *Meganeura*, which could achieve a wingspan of 75 cm (29½ in).

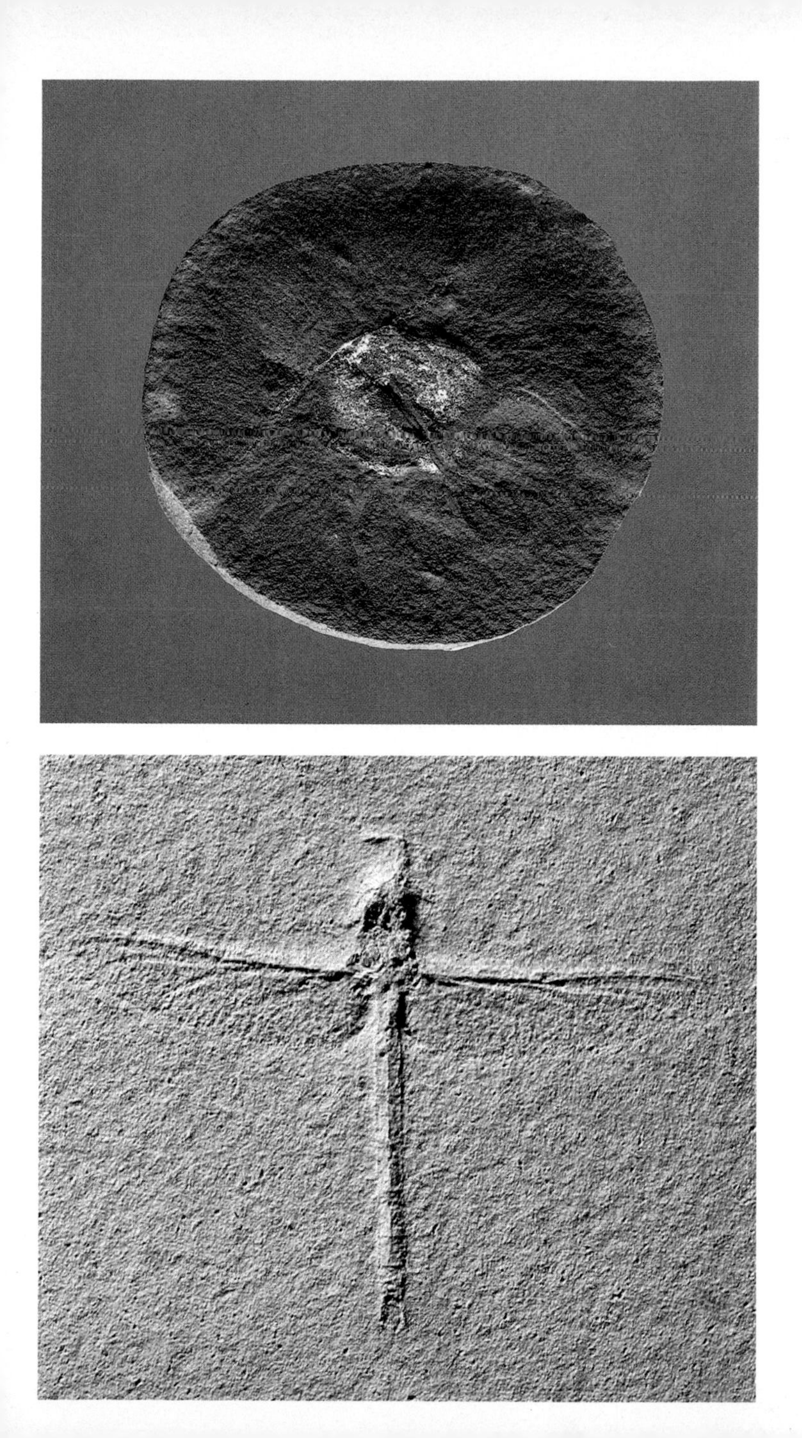

155 CRYPTOBLASTUS

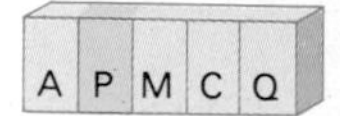

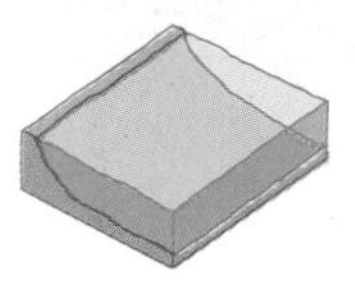

Classification Phylum Echinodermata, Class Blastoidea, Order Spiraculata, Family Granatocrinidae.
Description This is a representative of the Blastoidea, an extinct class of Echinodermata whose bodies were composed of a theca, or calyx, with small appendages, or brachioles, supported by an articulated stem. Blastoids had a respiratory structure (a hydrospire) within the theca, beneath the ambulacra; the latter were petal-shaped and bore brachioles. The theca, which in *Cryptoblastus* is elliptical and greater in height than in length, is made up of three series of plates: five deltoid plates surrounding the mouth, which is at the center of the upper region of the theca; five lateral plates, which are the more developed; and three basal plates further down. Around the mouth, which has a typical five-pointed star shape, are five pores. The ambulacral areas, which in fossils look dimpled because of brachial sutures, have central, longitudinally furrowed, calcareous structures called "lancet plates."
Stratigraphic position and geographical distribution The genus is typical of the Lower Carboniferous in the U.S. (Indiana and Montana). The examples in the photograph come from this area. They belong to the species *C. melo* and have a maximum size of 1 cm (½ in).
Note Blastoids, widespread from the Silurian to the Permian, occur in association with rugose corals, brachiopods, crinoids and bryozoans. Affixed to the sea floor, they ate plankton caught by their ciliated appendages.

---

156 HOLOCYSTITES

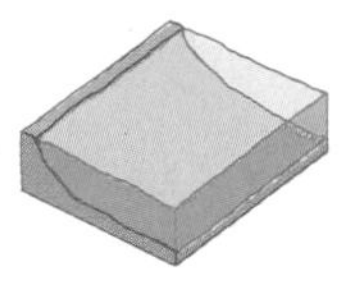

Classification Phylum Echinodermata, Class Cystoidea.
Description An echinoderm of the extinct class Cystoidea, characterized by a theca composed of numerous, random plates, which can vary in number from 13 to more than 2,000. In *Holocystites* there are about 100 of these plates, arranged in a dozen circlets. The theca, which in the genus illustrated measures up to 5 cm (2 in) in height and 4 cm (1½ in) in width and is roughly ovoid, is generally more or less spherical or sac-shaped in cystoids. One characteristic feature is the presence of short, unramified brachioles and pores that traverse the plates. On the basis of the nature of these pores, two orders of cystoids have been identified: one characterized by pores randomly arranged in pairs on the surface of the plates, and one characterized by the rhomb pore system, which takes the form of superficial grooves arranged in a rhombic pattern in the areas of suture between two contiguous plates.
Stratigraphic position and geographical distribution The genus is typical of the North American Silurian (Wisconsin, Illinois and Indiana). The example in the photograph comes from the Middle Silurian (420 million years ago) at Osgood, Indiana, and belongs to the species *H. scutellata*.
Note Like blastoids, cystoids were sessile organisms which affixed themselves to the sea floor, by a stem or directly, and lived by catching food with their brachioles. Numerous examples of *Holocystites* show holes perforating their surface plates: it is believed these were caused by gastropod parasites that fed on their skeletal tissue.

157 ABROTOCRINUS

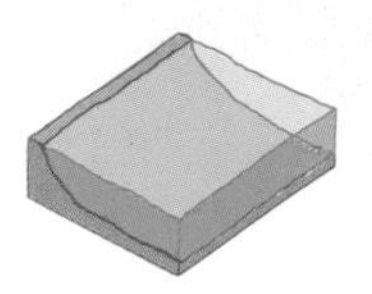

Classification Phylum Echinodermata, Class Crinoidea, Order Cladida, Family Staphylocrinidae.
Description A fine form of inadunate crinoid with long free arms above the radial plates. It consists of a shallow, cup-shaped theca, from which there emerges a crown formed of a bundle of elongate arms; the latter are circularly arranged in a single series and bifurcate at several points. The theca is composed of several series of plates: working upwards, there are five infrabasal plates, five radial plates larger than the preceding ones, and three anal plates. The theca or calyx is supported on a stem, which is of pentagonal section in the part lying nearest the theca and gradually becomes rounder as it diverges from it.
Stratigraphic position and geographical distribution The genus is restricted to the lower part of the Mississippian (Lower Carboniferous, 345 million years ago); it occurs in strata in the states of Illinois, Indiana, Iowa and Alabama. The example in the photograph, measuring approximately 10 cm (4 in), belongs to the type species *A. unicus* and comes from the Mississippian sediments of Crawfordsville, Indiana.
Note As in the majority of crinoids, the arms of *Abrotocrinus* develop in a single direction (upwards, in this case); this favors their preservation in a single bundle in the fossil state, as in the photograph.

---

158 HYPSELOCRINUS

Classification Phylum Echinodermata, Class Crinoidea, Order Cladida, Family Scytalocrinidae.
Description A crinoid belonging to the inadunate subclass, characterized by a calyx whose height is greater than its width and around which there is a crown of very long thin arms. The calyx is formed of elongate infrabasal plates, basal plates and broad, radial plates of polygonal form. The stem has a circular section.
Stratigraphic position and geographical distribution The genus is typical of the Lower Mississippian (the Mississippian corresponds approximately to the European Lower Carboniferous and began 345 million years ago). Its geographical distribution is limited to certain states of the U.S. The example in the photograph is from the Illinois Mississippian; it measures approximately 8 cm (3¼ in).
Note The genus forms part of a family of inadunate crinoids whose members were widespread in America from the Middle Paleozoic to the end of the era. They were all characterized by a tall, slender crown of mainly uniserial arms and a conical calyx.

159 ENCRINUS

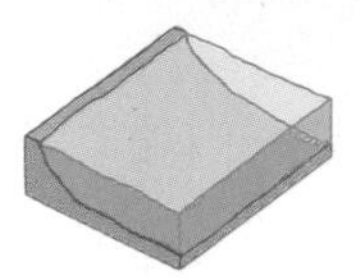

Classification Phylum Echinodermata, Class Crinoidea, Order Cladida, Family Encrinidae.
Description The calyx of this crinoid is only slightly developed in height and much more developed in width, with the radial and brachial plates arranged in three perfectly superimposed circles to form a beautiful, pentamerous symmetry. The arms, strong and equipped with pinnules, are plated at the base; they are very high and are characteristically arranged in a parallel alignment, very close to each other. The stem is elongate, stout and of circular section.
Stratigraphic position and geographical distribution The genus is very typical of the European Middle and Upper Triassic (220–205 million years ago). The example in the photograph belongs to the species *E. lilliformis* and comes from the German Middle Triassic; it measures approximately 7.5 cm (3 in).
Note *Encrinus* is often preserved in its entirety with its calyx and arms in the shelly limestone sediments (Muschelkalk) that characterize the central European Middle Triassic. The region was at that time covered by shallow seas in which crinoids were particularly abundant: to such an extent, in fact, that the remains of *Encrinus* and those of the related genus *Dadocrinus* are sometimes so plentiful that they become the main component of the rocks.

---

160 ONYCHOCRINUS

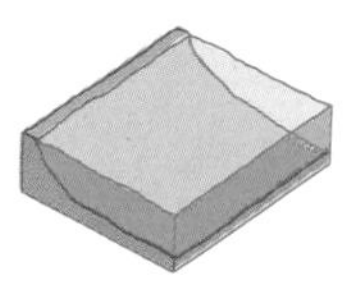

Classification Phylum Echinodermata, Class Crinoidea, Order Taxocrinida, Family Synerocrinidae.
Description A beautifully preserved specimen of Carboniferous crinoid characterized by a very low crown made up of stumpy arms borne on five main rays. The arms are repeatedly ramified to form further, short arms. The stem is elongate, of circular section and broadens near the calyx.
Stratigraphic position and geographical distribution The genus is known in the Lower Carboniferous (345–315 million years ago). It occurs in Europe (Ireland, Germany, Scotland), Canada and certain parts of the U.S. (Illinois, Indiana, Iowa, Alabama, Oklahoma). The example in the photograph, *O. exculptus*, comes from the Lower Carboniferous (Mississippian) at Crawfordsville, Indiana.
Note *Onychocrinus* belongs to a crinoid subclass, the Flexibilia, which are derived from the inadunate crinoids. The Flexibilia form a well-characterized group of Paleozoic crinoids whose peculiar features consist of uniserial arms lacking pinnules and often inward curling, as can be seen in *Onychocrinus*, and a stem with no cirri; the order Taxocrinida is also equipped with an anal tube.

161 PENTACRINITES

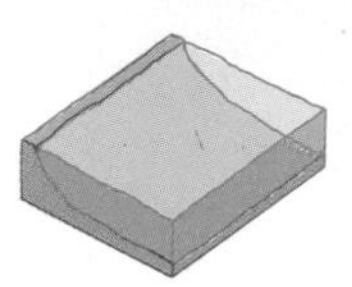

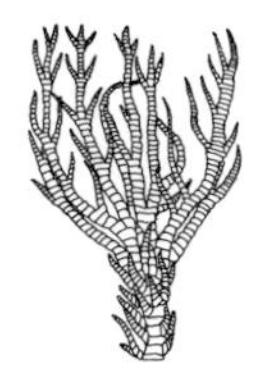

Classification Phylum Echinodermata, Class Crinoidea, Order Isocrinida, Family Pentacrinitidae.
Description A characteristic Mesozoic crinoid, which is often referred to as *Pentacrinus*. It may include forms of considerable size, with a stem up to one meter (39 in) long, equipped with numerous highly developed cirri, which extend upwards and completely envelop the stem and the calyx. In section, the calyx is either lobate or shaped like a pentagonal star. The calyx, very small and composed of two series of plates, supports elongate arms which are extensively ramified and equipped with numerous pinnules.
Stratigraphic position and geographical distribution The genus ranges from the Lower to the Upper Jurassic (190–145 million years ago) and is characteristic of central and northern Europe, but it also occurs in North America. The example in the photograph comes from the French Jurassic and as can be seen, stellate elements and cirri, scattered through the sediment after the death of the organism, sometimes represent an important component of the rock.
Note Isocrinida still live in the oceans of the world today, where they form large meadows on the sea floor. In the Lower Jurassic, *Pentacrinites* and related forms such as *Seirocrinus*, up to 20 meters (6½ ft) long, enjoyed a period of considerable development. Famous remains of these large crinoids have been discovered, entire and still attached to the floating tree trunks (similarly fossilized) to which they affixed themselves, in the Holzmaden layer in southern Germany.

---

162 PTEROCOMA

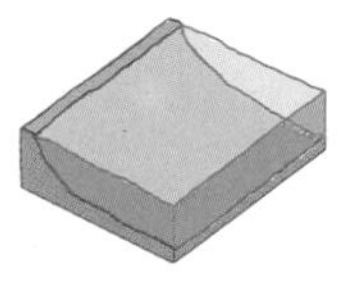

Classification Phylum Echinodermata, Class Crinoidea, Order Comatulida, Family Pterocomidae.
Description A stemless crinoid, with a central disk of very small diameter (a few millimeters) and thin, elongate arms some 12 cm (5 in) long. The arms bear numerous elongate pinnules.
Stratigraphic position and geographical distribution The genus ranges from the Upper Jurassic to the Turonian stage of the Upper Cretaceous (140–90 million years ago). It occurs in the Upper Jurassic (Kimmeridgian) at Solnhofen (Germany), and in the Lebanese Upper Cretaceous. The example in the photograph, of the species *P. pinnulata*, comes from the Lebanese Upper Cretaceous at Sahel Alma; it measures approximately 5.5 cm (2¼ in).
Note The genus is very similar to the living *Antedon*, a comatulid crinoid which in its adult stage generally lives in groups among seaweed. The fossil genus *Pterocoma* must also have lived in groups, indicated by the discovery of slabs containing several specimens. It was a planktonic form that floated with the currents.

163 SALTERASTER

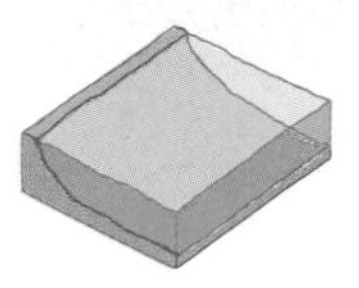

Classification Phylum Echinodermata, Class Stelleroidea, Subclass Asteroidea, Family Urasterellidae.
Description This is a starfish dating back to the Paleozoic, characterized by five arms that give it its typical pentamerous symmetry. The arms are very broad and strong, with a circular section; they possess a single row of median dorsal plates, separated from the marginal plate by numerous papillae. Some of these (adambulacral) plates, situated on the lower surface of the arm, bear transversal furrows with robust spines. The madreporite is positioned on the dorsal region of the central disk; in echinoderms, this plate acts as a valve to make water flow into the internal vascular system.
Stratigraphic position and geographical distribution *Salteraster* ranges from the Middle Ordovician to the Silurian (465–395 million years ago); it occurs in North America, England and Australia. The slab in the photograph comes from the Australian Silurian; the examples are approximately 4 cm (1½ in) wide.
Note It is probable that *Salteraster*, like some modern starfish, fed on animals with a shell, opening the valves of its prey by means of the podia with which its arms were equipped, and everting its stomach; the digestion of its prey therefore took place outside the body. In order to carry out this operation the animal had to stand over its prey, raising itself up on its arms; there is an example of a fossil preserved in this position.

---

164 OPHIOPINNA

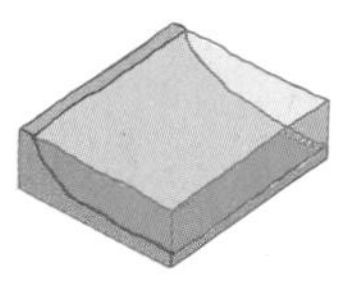

Classification Phylum Echinodermata, Class Stelleroidea, Subclass Ophiuroidea, Family Ophiacanthidae.
Description This is an ophiuroid very similar to many living forms; the body is composed of a central disk covered in thin scales. The arms are very long in comparison to the disk, broad in the center and thinning down as they approach the tip. A few rows of spines occur at the initial part of the arms.
Stratigraphic position and geographical distribution The genus ranges from the Pliensbachian to the Callovian (Lower and Middle Jurassic, 180–160 million years ago). Its geographical distribution is somewhat limited, found solely in French and Swiss Jurassic sediments. The slab in the photograph comes from the French Jurassic; the examples measure approximately 4 cm (1½ in) at most.
Note The general morphology of the ophiuroid body has in many cases remained constant right up to the present, even though the earliest representatives of the group date back to the Ordovician. The skeletal structure of the arms, however, has greatly evolved: their plates, or vertebrae, have adopted a conformation that allows for movements reminiscent of those made by snakes, and it is to this method of arm movement that ophiuroids owe their name. The vertebrae of ophiuroids consist of disks derived from the fusion of four ambulacral plates and are interarticulated.

165 TAENIASTER

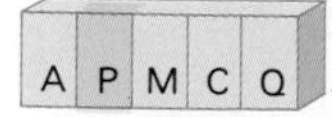

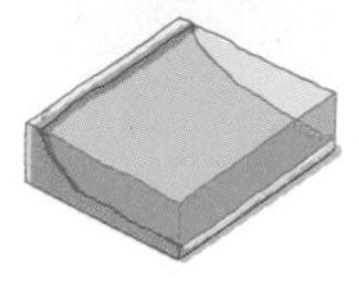

Classification Phylum Echinodermata, Class Stelleroidea, Subclass Ophiuroidea, Family Protasteridae.
Description An early form of ophiuroid, characterized by elongate and fairly robust arms, with a clearly marked longitudinal depression between the ambulacral plates. The side plates of the arms are folded back so as to form a sort of lateral protection. The central disk is large, covered in scales and not clearly defined at the margins.
Stratigraphic position and geographical distribution The genus ranges from the Middle Ordovician to the Lower Devonian (465–380 million years ago); it is found in North America and Germany. The example in the photograph comes from the German Lower Devonian at Bundenbach and measures approximately 10 cm (4 in).
Note The fossil remains of ophiuroids, although not frequent, are more numerous than those of starfish. These remains often consist of examples preserved *in toto*, with the different pieces of the skeleton anatomically connected rather than scattered through the sediment. An excellent example of preservation is provided by the extremely fine pyritized remains of *Taeniaster* discovered in the black shale rocks of Bundenbach, known as Hunsrückschiefer. X-rays are used to examine these fossils: in radiography, the pyrite that has impregnated the fossil brings out the details of the tiniest structures in a way that would otherwise be impossible.

---

166 FURCASTER

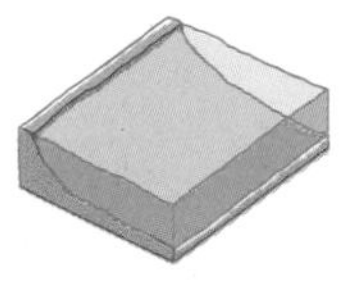

Classification Phylum Echinodermata, Class Stelleroidea, Subclass Ophiuroidea, Family Furcasteridae.
Description Another early form of ophiuroid, characterized by a broad disk and five robust arms, which, in ophiuroids, contrary to those of starfish, are incapable of regeneration once they have become detached from the body. These arms are equipped with needle-shaped spines, all more or less of the same dimensions, arranged in rows near the sides of the arms and parallel to their axis.
Stratigraphic position and geographical distribution The genus ranges from the Upper Ordovician to the Lower Carboniferous (450–325 million years ago). It occurs in some European layers and in North America and Australia. The slab in the photograph comes from the German Devonian at Bundenbach; the examples measure approximately 5 cm (2 in.).
Note As in the case of *Taeniaster*, some extremely fine and complete specimens of *Furcaster* occur in the black shale rocks of Bundenbach (Germany). These fossils often show the arms of the various specimens all facing in the same direction, as in the photograph. This is due to the existence of currents near the ancient seabed, which imposed the same orientation on the remains; the fossils thus record the existence and directions of the undercurrents present at the moment of their deposition.

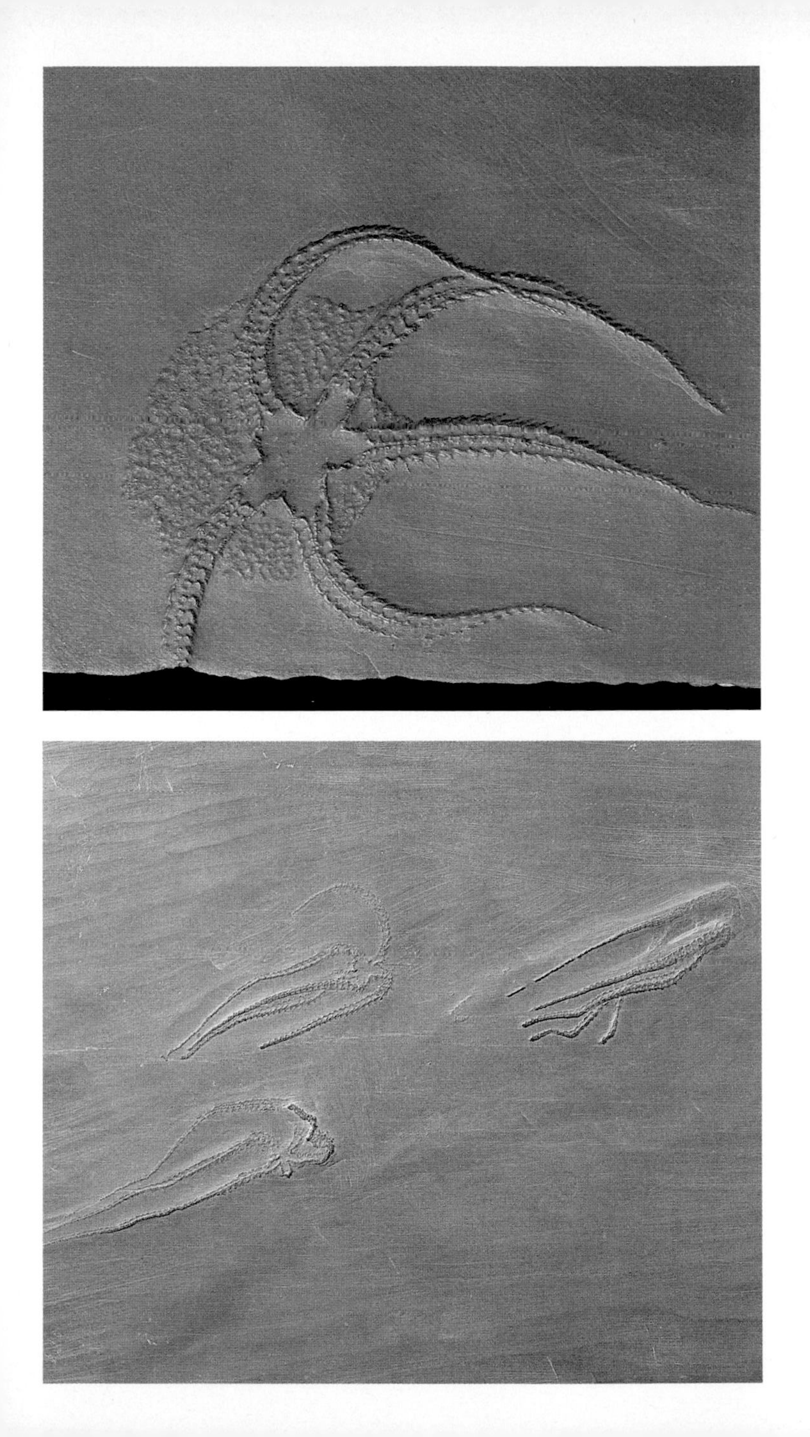

167 CLYPEASTER

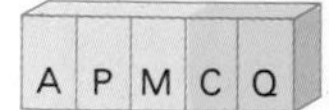

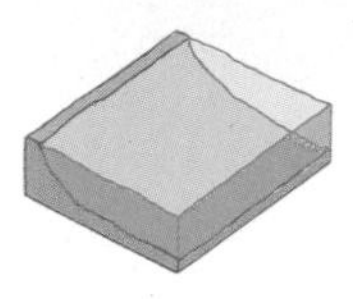

Classification Phylum Echinodermata, Class Echinoidea, Order Clypeasteroida, Family Clypeasteridae.
Description Echinoid of medium to large dimensions, with a test of pentagonal outline and ornamented with granules, which in some forms displays a raised, bell-shaped surface in the central area; the margins are either rounded or flattened. The upper surface has ambulacra of a distinct, rounded petaloid shape; the shape of these petals is emphasized by the presence of pores. The anal aperture of the test, or periproct, is located on the oral side near the posterior margin and, in some forms, actually on the margin. The oral side of the test, where the mouth occurs, may be either flattened or convex. The posterior jaw is of greater size than the others. The general morphology of the test and the shape of the petals does, nevertheless, vary a great deal in the numerous species contained within this genus.
Stratigraphic position and geographical distribution The earliest representatives of this genus date from the Upper Eocene (40 million years ago) at Antwerp (Belgium), and the genus is still widespread in seas today. The example in the photograph comes from the Italian Pliocene in Calabria and measures approximately 15 cm (6 in).
Note Some species of *Clypeaster* are widely used in the stratigraphy of Tertiary rocks. *Clypeaster* favor warm, turbulent waters. Some species live on sandy sea floors, burrowing well beneath the surface of the sediment.

---

168 DENDRASTER

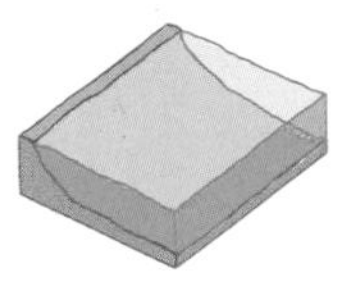

Classification Phylum Echinodermata, Class Echinoidea, Order Clypeasteroida, Family Dendrasteridae.
Description Echinoid, either medium or large in size, with a very flattened test and a rounded outline. The apical system, which occurs at the center of the pattern made by the ambulacra, lies toward the posterior. The ambulacra are petaloid and rounded: the posterior ones are short while the anterior one is much longer. The margins are tapered. The periproct (anal aperture) is positioned on the oral surface near the margin.
Stratigraphic position and geographical distribution This is a recent genus, which first appeared in the Pliocene (5 million years ago) and is still widespread off the west coast of America. The example in the photograph, which measures approximately 8 cm (3¼ in) wide, comes from the Pliocene in the Kettleman Hills, California, and belongs to the species *D. coalingensis*.
Note These flattened echinoids, known in the U.S. as "sand dollars," are gregarious organisms living in such numerous groups that in some Pliocene sediments in California they represent the main component of the local rock. Some modern species live in sandy environments at a very oblique angle to the sediment, burying only the anterior portion of their test beneath the sand and leaving the remainder protruding from it.

169 HEMIPNEUSTES

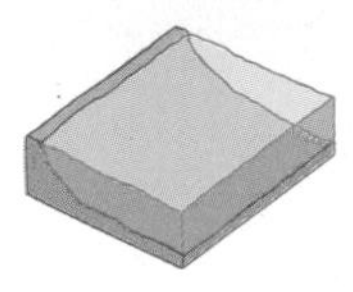

Classification Phylum Echinodermata, Class Echinoidea, Order Holasteroida, Family Holasteridae.
Description The test of this sea urchin is ovoid in outline, with a smooth, very convex upper surface; the anterior ambulacrum (third ambulacrum) is sunken in an elongate depression in the test. The ambulacra are small and petaloid. The peristome (the aperture in the shell surrounding the mouth) is anteriorly sited and is in the shape of a half moon with rounded extremities. The periproct is situated below the margin.
Stratigraphic position and geographical distribution The genus occurs in the Upper Cretaceous, from the Santonian to the Campanian stages (82–68 million years ago). Its geographical distribution is restricted to northern Europe. The example in the photograph, approximately 8 cm (3¼ in) long, comes from the Dutch Upper Cretaceous in Limburg and belongs to the type species *H. striatoradiatus*.
Note This order is composed of echinoids whose earliest representatives date back to the Lower Jurassic, but includes living forms. Of the genera classified in the family Holasteridae, only one contains forms still living: the genus *Stereopneustes*, a sea urchin widespread in Japan and the Indian Ocean, characterized by a very thin test and living at depths between 250 and 900 meters (2,950 ft).

---

170 STENONASTER

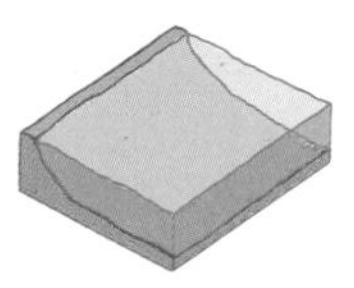

Classification Phylum Echinodermata, Class Echinoidea, Order Holasteroida, Family Stenonasteridae.
Description A typical echinoid form, with the test aborally conical, orally flat and with an irregular outline. The ambulacra are radially arranged and are not petaloid; their pores are of elongate shape. The periproct is well developed and is situated inframarginally (on the oral side, near the margin). The mouth peristome lies in a forward position and is of rounded half moon shape. The aboral side of the test is characterized by markedly convex plates creating the general impression of an ornament made up of thick tubercles.
Stratigraphic position and geographical distribution This is a typical genus of the Mediterranean Upper Cretaceous (80 million years ago); it occurs particularly in Cretaceous sediments in Italy and Tunisia. The example in the photograph, which belongs to the type species *S. tubercolata*, comes from the Senonian (88–65 million years ago) in the foothills of the Alps near the Veneto region in Italy and is approximately 4 cm (1½ in) long.
Note The species *S. tubercolata* is used as guide fossil for the Italian Senonian in the Veneto and the central Apennines. The genus *Stenonaster*, for a long time also referred to by scholars as *Stenonia*, is named after Nicholas Steno, the 17th-century Danish scholar who became one of the precursors of modern scientific geological studies.

171 MICRASTER

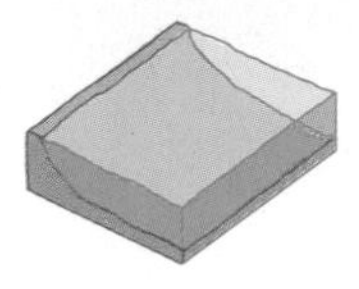

Classification Phylum Echinodermata, Class Echinoidea, Order Spatangoida, Family Micrasteridae.
Description A typical form of fossil echinoid, with a globular test whose margins have a characteristic heart shape owing to the presence of an anterior groove and a contraction in the test at the point where the anus is sited. The petaloid ambulacra are broad and clearly marked, and the surface of the test is slightly granular in appearance. The mouth is anteriorly situated on the ventral side, also granular, and is covered by a fold in the test that forms a lip (labrum). The posterior region, where the anus is located, is vertical. Like other Spatangoid echinoids, *Micraster* lacks masticatory apparatus.
Stratigraphic position and geographical distribution The genus ranges from the Upper Cretaceous (Cenomanian stage) to the Danian stage of the Lower Paleocene (100–63 million years ago). It occurs in various European layers, in Malagasy and in Cuba. The example in the photograph, approximately 6 cm (2½ in) long, belongs to the type species *M. coranguinum* and comes from the French Upper Cretaceous at Meudon.
Note Some species of *Micraster*, such as *M. coranguinum*, are used as guide fossils in stratigraphy. A classic example of evolutionary succession is provided by the different forms of *Micraster* occurring in English (Cretaceous) chalk beds, which reveal the progressive adaptation of these sea urchins to life buried in loose marine sediments.

---

172 EUPATAGUS

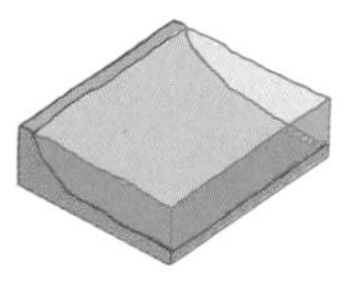

Classification Phylum Echinodermata, Class Echinoidea, Order Spatangoida, Family Brissidae.
Description A living sea urchin, possessing a low, ovoid shell with a flattened oral surface. It is a form with fascioles, narrow bands densely covered in ciliate spines designed to create currents in the water; these ciliate bands can be detected on tests deprived of their spines by the rows of tubercles that support the spines. The test possesses an apical system, made up of plates positioned at the dorsal end of the ambulacra and interambulacra, which is anteriorly displaced. The ambulacra, of equal sizes, are petaloid. The upper or "aboral" surface reveals tubercles matching the fascioles.
Stratigraphic system and geographical distribution The genus first appears in the Eocene (55 million years ago) and includes forms living in seas today. The example in the photograph, approximately 5 cm (2 in) long, belongs to the species *E. antillarum* and comes from the Florida Eocene.
Note Spatangoid echinoids such as *Eupatagus* are specially adapted to living in tunnels burrowed in muddy or sandy sea-beds, even though some forms simply rest on the sea floor. To solve the problems inherent in this lifestyle, Spatangoida have characteristic adaptive features such as a depression in the test, which encourages the channeling of currents of water toward the respiratory structures, or the development of fascioles.

173 ACHISTRUM

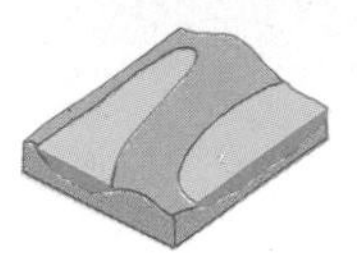

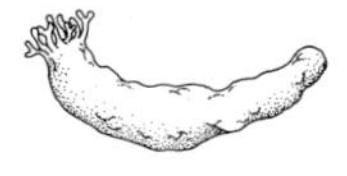

Classification Phylum Echinodermata, Class Holothuroidea.
Description This is a holothuroid or sea cucumber, an echinoderm that differs markedly from the other members of the same phylum in that its body, of circular section, elongate and similar to that of a large worm, is not covered by a skeleton of calcareous plates, but is equipped with tiny calcite sclerites of varying shape situated within the dermis. The mouth is at one end of the vermiform body, the anus at the other. The animal always appears flattened inside the siderite nodules within which it is preserved, with rounded anterior and posterior extremities.
Stratigraphic position and geographical distribution The genus *Achistrum*, containing the single species *A. welleri*, is known in the Mazon Creek beds in Illinois, dating from the Pennsylvanian (Upper Carboniferous, 300 million years ago); the example in the photograph measures approximately 8 cm (3¼ in) in length.
Note Fossil holothurians are known from the Middle Cambrian, although they occur infrequently in fossil form. Living holothurians are exclusively marine organisms that live for the most part on the sea floor, where they feed by ingesting large quantities of mud, from which they absorb organic particles, expelling what they are unable to digest. Living holothurians display even more highly advanced regenerative capabilities than starfish when broken into two or three pieces.

---

174 ARCHISYMPLECTES

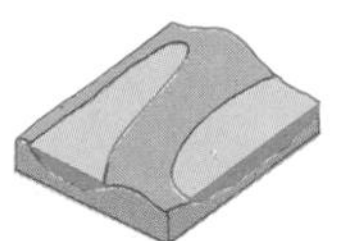

Classification Phylum Nemertinea.
Description The remains of *Archisymplectes* consist of a simple film or trace of color, in the shape of a narrow band and characteristically multiplicate; some fossil examples retain the extrovertible proboscis with which these organisms are endowed; this proboscis, often equipped with a hard point, acts as a means of paralyzing or killing prey. Since these animals had no parts, nemertine remains are extremely rare in fossil form.
Stratigraphic position and geographical distribution The genus is found in the Upper Mississippian at Bear Gulch, Montana, and in the Upper Pennsylvanian at Mazon Creek, Illinois (Lower to Upper Carboniferous, 325–280 million years ago). The example in the photograph belongs to the only known species, *A. rhothon*, and comes from Mazon Creek.
Note Nemertinea are predominantly marine worms; they live for the most part close to the coast, hidden in the mud or beneath pebbles, or in the open sea; few species inhabit fresh water. They are of modest length (about 1 centimeter or ½ in), but there are forms, such as the North Sea *Lineus*, which reach 20 and even 30 meters (65½–98½ ft) in length. Nemertine worms are carnivorous predators.

175 COPRINOSCOLEX

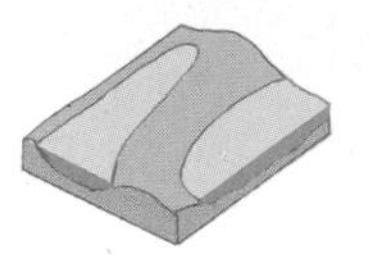

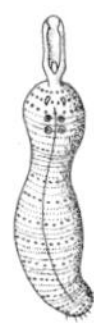

Classification Phylum Echiurida.
Description A highly unusual type of fossil. It is the remains of a vermiform organism of roughly ovoid shape, elongate, covered externally by an annulate cuticle and with a proboscis at the anterior extremity. Because of the absence of hard parts and the simplicity of the living organism's form, few observations can be made on the fossil specimens. They do nevertheless represent an example of exceptional fossilization.
Stratigraphic position and geographical distribution The genus *Coprinoscolex* is known in a single species, *C. ellogimus*, from Mazon Creek, Illinois (Pennsylvanian, Upper Carboniferous, 300 million years ago). The example in the photograph, which measures approximately 3 cm (1¼ in) comes from Mazon Creek.
Note Echiurida represents a small group of vermiform organisms whose bodies are equipped with a non-retractable proboscis, which at present contains 150 living species. They are exclusively marine, living nestled in sand or mud, or hidden in rock clefts in shallow waters. Abyssal forms also occur.

---

176 SPRIGGINA

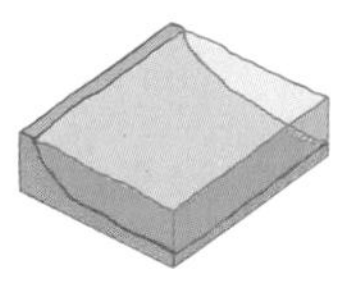

Classification Phylum Annelida, Class Polychaeta, Family Sprigginidae.
Description An organism with an elongate, vermiform body on which are clearly visible a head, a segmented trunk, which may contain up to 42 segments, and a small pygidium in the terminal portion. One particularly characteristic feature is the head, which is horseshoe-shaped—narrow and very curved—with two pointed lateral projections at the back. The body is crossed longitudinally by a central groove corresponding, in all probability, to the digestive apparatus, and by two lateral grooves corresponding to the longitudinal musculature. The sides of the segments sometimes bear the imprints of small, needle-shaped setae.
Stratigraphic position and geographical distribution *Spriggina* includes the species *S. floundersi*, an example of which is shown in the photograph, found in the Australian Precambrian at Ediacara (670 million years ago). Examples similar to *Spriggina* have recently been identified in Precambrian rocks in southern Africa and the northern U.S.S.R.
Note *Spriggina* is one of the most characteristic fossil forms of the Ediacaran Precambrian fauna. The original environment in which these fossils were laid down must have been shallow, littoral waters that were warm, with normal salinity and oxygenation. Apart from the forms illustrated in this book, the Ediacaran fauna includes other coelenterates, arthropods and organisms of unknown zoological affinities.

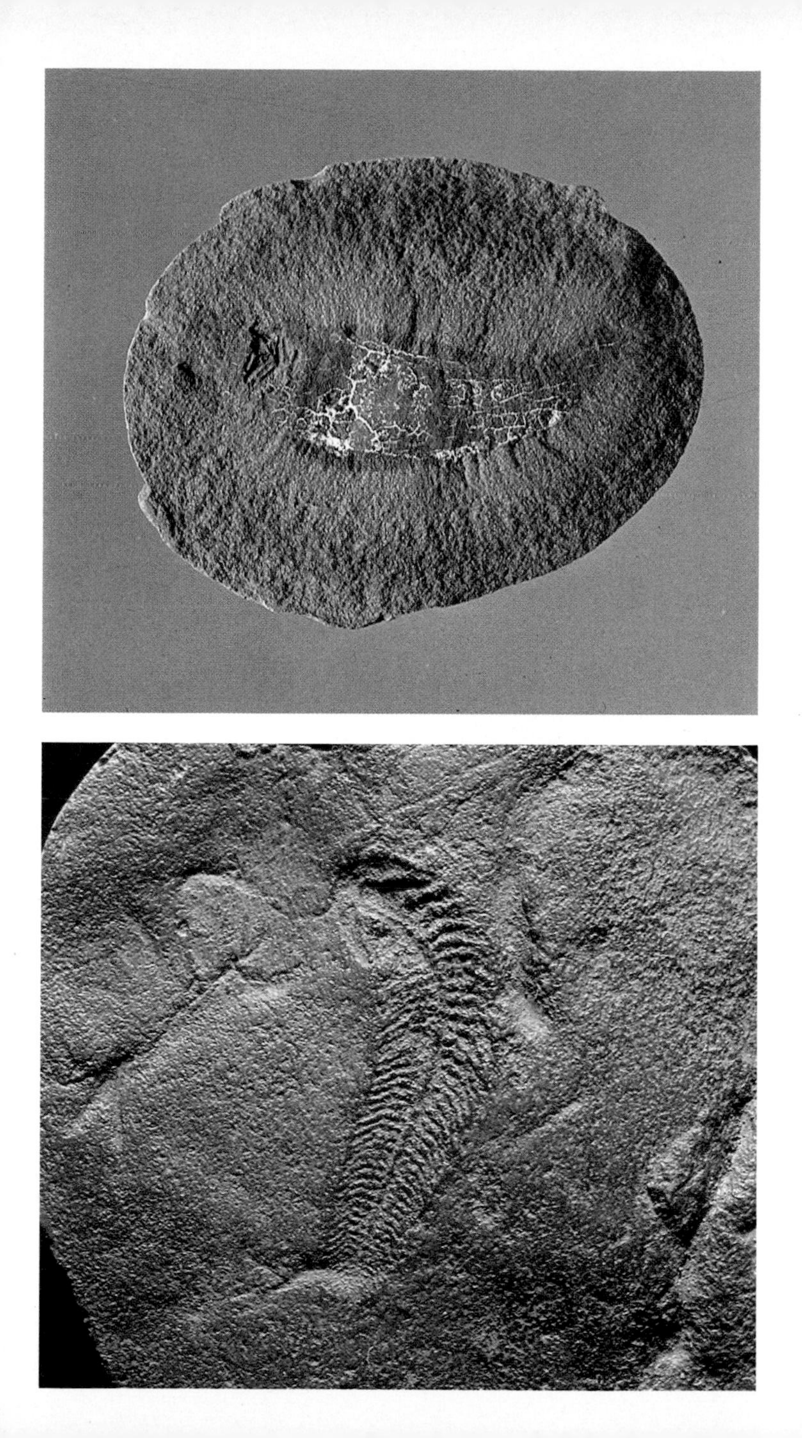

177 DICKINSONIA

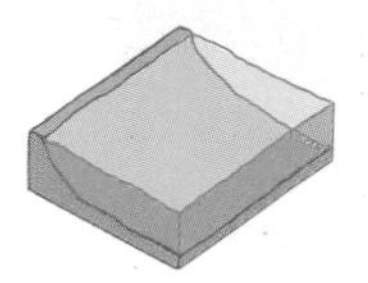

Classification Phylum Annelida, Class Polychaeta, Family Dickinsonidae.
Description Ancient polychaete annelid with a broad, flat body of considerable size: up to 45 cm (18 in) long in the largest known specimen. The outline of the body is more or less elliptical, with subparallel sides; it is marked by a central, longitudinal axis, corresponding to the intestine, and is divided into numerous segments that form a dense transversal design; the segments in the region in front of the mouth are fused together. Dorsal lamellae have been preserved on some specimens.
Stratigraphic position and geographical distribution The genus is represented by a number of different species from the Upper Precambrian (circa 670 million years ago) in the Ediacara Hills, South Australia. The example in the photograph is the type species *D. costata*.
Note This is an animal that lived close to the sea floor, in the shallow waters of the ancient Ediacara Sea. It procured food by filtering the detritus on the seabed in search of organic particles and perhaps possessed the ability to swim freely. There are surprising similarities between this primitive annelid and the modern polychaete *Spinther*, from which it is separated by a timespan of almost 700 million years.

---

178 BURGESSOCHAETA

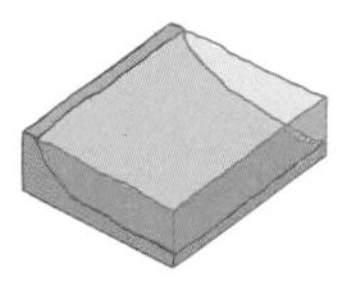

Classification Phylum Annelida, Class Polychaeta.
Description A small polychaete annelid, a type characterized by the presence of numerous bristles (*chaite* in the ancient Greek), with an average body length of approximately 3 cm (1¼ in). The body is subdivided into segments, whose number may exceed 20 in adult organisms. On the anterior part of the body there are two flexible, elongate tentacles, and the animal may also possess an eversible proboscis. Each segment of the body (except the first) is characterized by the presence of a couple of parapodia (appendages with a locomotory function) on each of its sides. The parapodia consist of an upper notopodium and a lower neuropodium and are easily distinguishable in fossils because they bear a tuft of long bristles (setae); the first segment bears only a couple of tufts.
Stratigraphic position and geographical distribution *Burgessochaeta* contains a single species, *B. setigera*, known solely from the Middle Cambrian Burgess Shale (circa 530 million years ago) of British Columbia, famous for its wealth of soft-bodied fossils such as polychaetes.
Note *Burgessochaeta* must have been a worm that lived in sediments, moving out of its tunnels by using its bristles and acquiring food, in the form of minute particles, by sticking its tentacles out from beneath the sediment.

179 HYSTRICIOLA

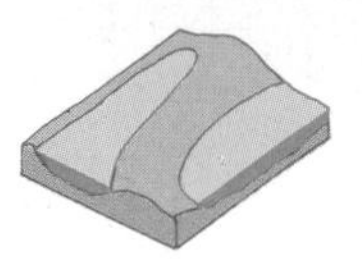

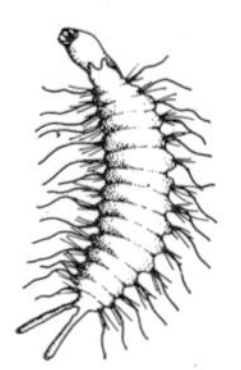

Classification Phylum Annelida, Class Polychaeta, Order Phyllodocida, Family Aphroditidae.
Description Annelid with a large, flat body, normally small in size (from 8 to 9 mm or just over ¼ in), with an average length of approximately 17 mm (.5–.75 in), subdivided according to size into segments ranging in number from ten to 24. The profile is ovoid. In the anterior region there are two antennae, while in the posterior region the last segment bears two long (anal) cirri. One characteristic feature are the two long parapodial cirri, which number one per parapodium. The bristles form a tuft on each parapodium, each tuft being accompanied by an aciculum. In the cephalic region of some examples it is possible to make out an eversible proboscis bearing at least four denticles, triangular in shape, bilobate at the base and hooked on the upper margin. The examples of *Hystriciola* take the form of light impressions in the rock, sometimes recognizable only through the difference in color of the fossil when compared to the rock.
Stratigraphic position and geographical distribution The genus has a single species, *H. delicatula*, found in the Pennsylvanian (Upper Carboniferous, 300 million years ago) at Mazon Creek, Illinois, where the example comes from.
Note The genus has been attributed to the living family Aphroditidae because of the general form of its body. It must have lived in piles of vegetable remains, both for protection and in order to gather food such as tiny crustaceans and other minuscule organisms (from 8 to 9 mm or just over ¼ in).

---

180 DRYPTOSCOLEX

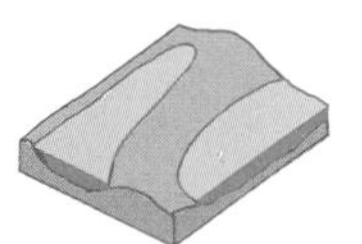

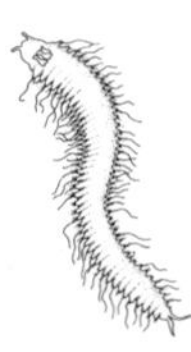

Classification Phylum Annelida, Class Polychaeta, Order Phyllodocida, Family Aphroditidae.
Description Polychaetes of small size, with a vermiform body reaching 122 mm (4½ in), subdivided into more than 150 segments. On the cephalic region it is sometimes possible to detect two short antennae, which do not exceed 4 mm (less than ¼ in) in length. In the majority of fossil examples an eversible proboscis is visible, armed with four mandibles. This proboscis, however, is rarely seen everted from the oral cavity, just as all four mandibles are rarely found preserved on the same specimen. The parapodia are biramous (divided into upper and lower parts), with each bearing a dense bundle of long thin bristles. There are long irregular cirri present on the parapodia and two anal cirri, which are rarely preserved in fossils.
Stratigraphic position and geographical distribution The genus has one species, *D. matthiesae*, which is found occasionally in the siderite nodules of Mazon Creek, Illinois, dating from the Pennsylvanian (Upper Carboniferous, 300 million years ago).
Note *Dryptoscolex* was an active predator that hunted its prey while swimming freely through the water. Like *Hystriciola*, it belongs to the living family Aphroditidae.

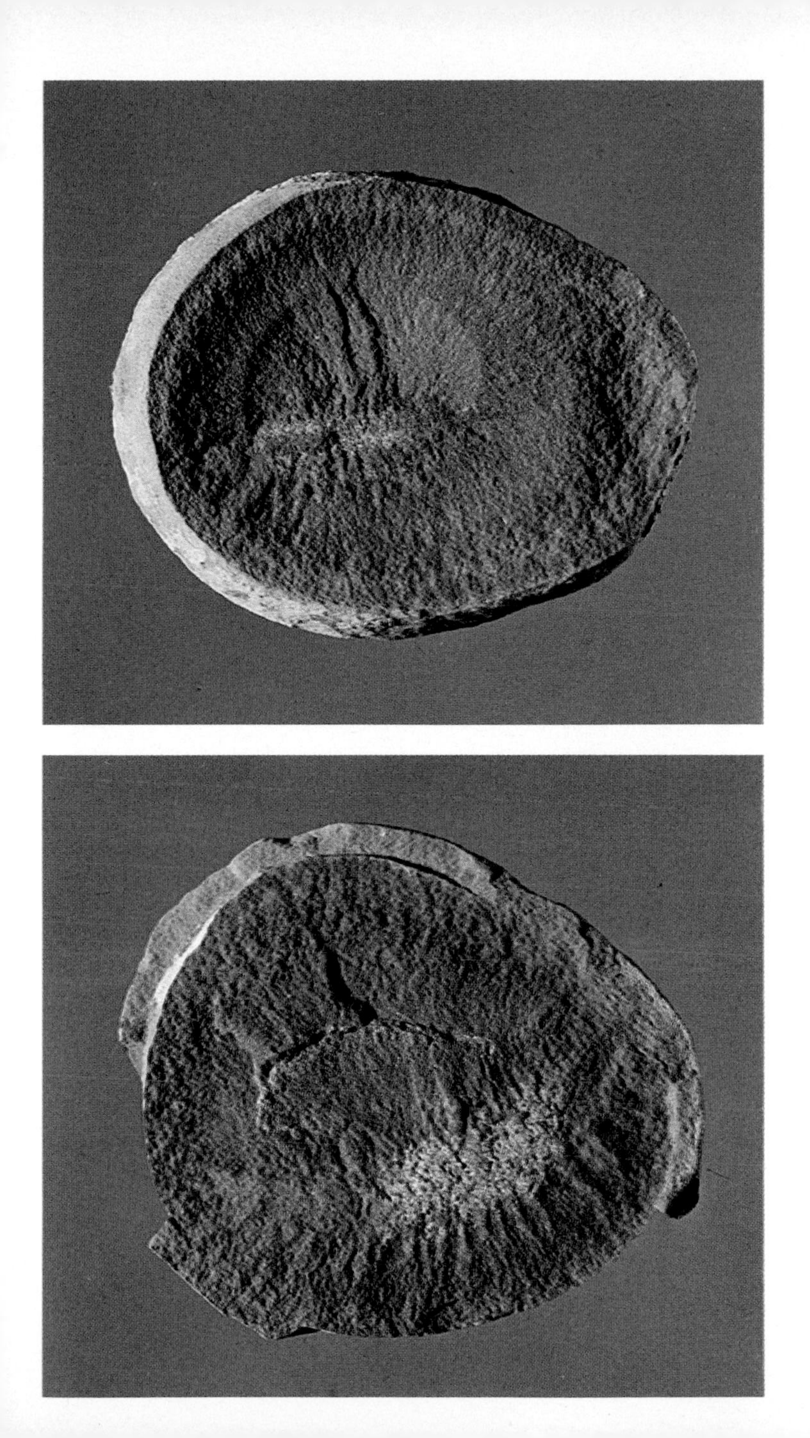

181 ASTREPTOSCOLEX

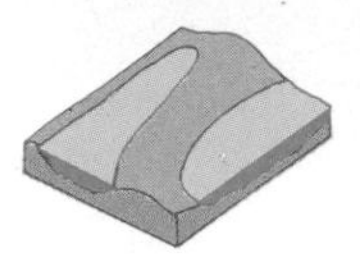

Classification Phylum Annelida, Class Polychaeta, Order Phyllodocida, Family Nephthydidae.
Description This polychaete has an elongate, robust, vermiform body, subdivided into 29 to 42 segments. The total length of the animal can vary from 16 to 55 mm (½–2 in). The head is rounded with two short antennae; it is also possible to detect the presence of an eversible proboscis and two jaws of roughly conical shape. The parapodia are equipped with clusters of short bristles: these are biramous and in certain cases they also bear comb-shaped gills. The small spines are often preserved and number one or two on each branch of the parapodia. The body tapers in the posterior region, which bears two long anal cirri at its tip.
Stratigraphical position and geographical distribution The genus is known solely in the Mississippian in Montana and the Pennsylvanian in Illinois (Lower and Upper Carboniferous, 360–300 million years ago). The example in the photograph, which belongs to the species *A. anasillosus*, comes from the Pennsylvanian of Mazon Creek, Illinois.
Note *Astreptoscolex* belongs to a family that includes numerous living forms; these live mainly in sand or mud on the sea floor, where they dig burrows with the help of their proboscis. *Astreptoscolex* must have been a carnivorous or omnivorous organism.

---

182 DIDONTOGASTER

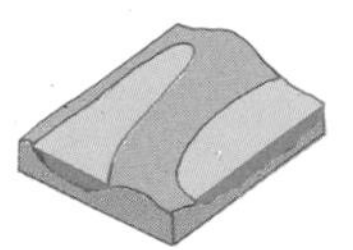

Classification Phylum Annelida, Class Polychaeta, Order Phyllodocida, Family Nephthydidae.
Description A polychaete of modest size whose average length never exceeds some 5 cm (2 in). The body is subdivided into numerous segments and is more elongate and densely segmented in its anterior section, tapering visibly in the posterior region. The head bears no type of appendage; there is a proboscis that can extend out of the oral cavity, which is equipped with two conical mandibles. The parapodia are well developed and biramous, with one upper and one lower branch, sustained internally by a small spine; they also bear tufts of numerous, rather short bristles.
Stratigraphic position and geographical distribution The genus has a single species, *D. cordylina*, found in the Pennsylvanian (Upper Carboniferous, 300 million years ago) of Mazon Creek, Illinois, where it occurs with sufficient frequency to make it the most abundant form of polychaete in these beds, famous for its numerous soft-bodied forms.
Note On the basis of the stomach contents preserved in some specimens, it has been deduced that *Didontogaster* was an omnivorous worm, feeding both on small invertebrates and particles of vegetable matter.

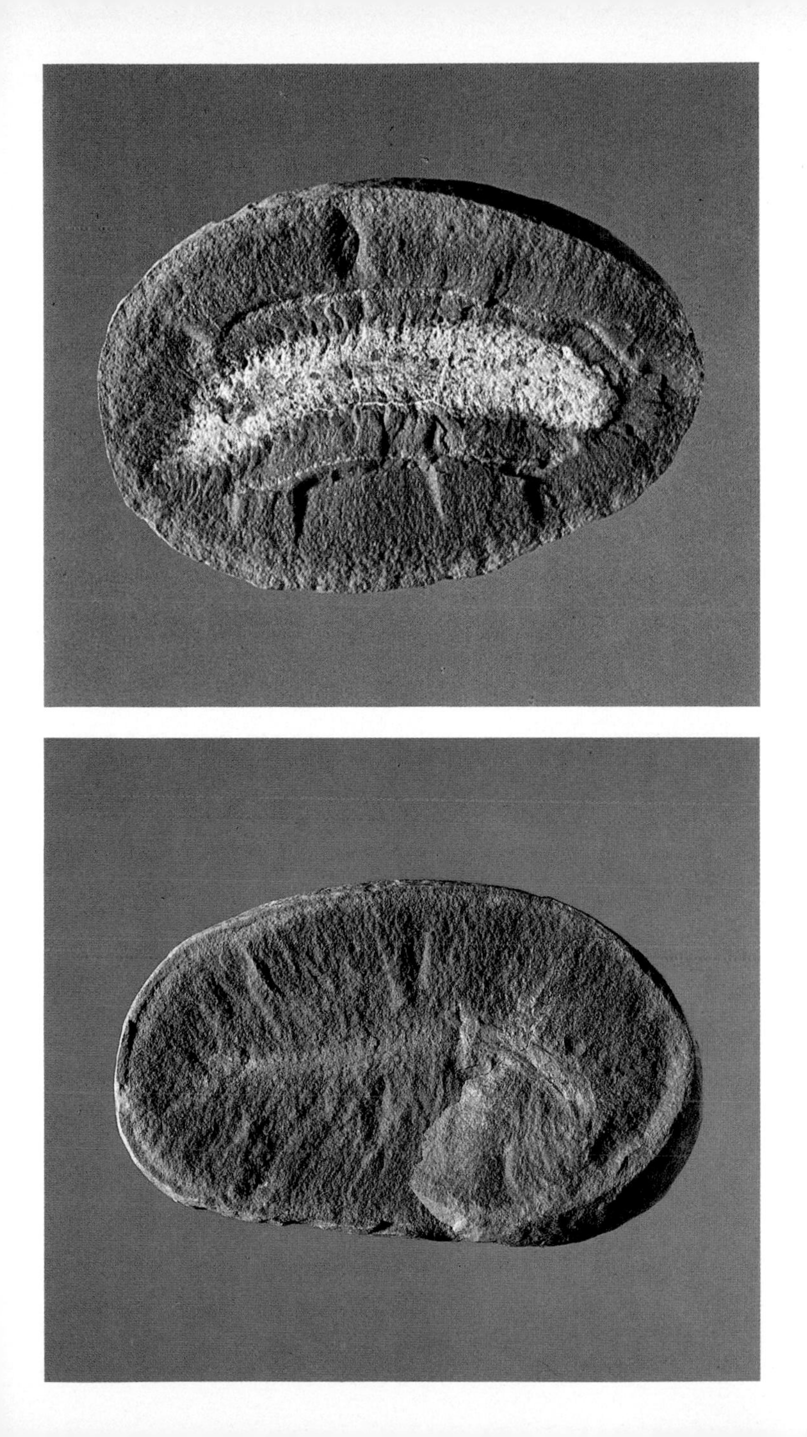

183 ESCONITES

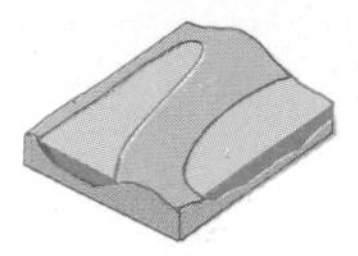

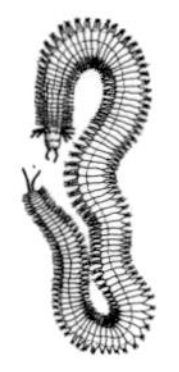

Classification Phylum Annelida, Class Polychaeta, Order Eunicida, Family Eunicidae.
Description A polychaete with a vermiform body that can grow to 14 cm (5½ in) and which is subdivided into as many as 80 segments. On the head there are numerous appendages, consisting of two palps and five antennae. One characteristic feature of this animal is the mouth apparatus, comprising two wing-shaped mandibles and maxillae composed of numerous chitinous plates, which together form a clearly recognizable complex in fossil specimens. The parapodia are biramous. In the anterior part of the body the upper branch of each parapodium bears comb-shaped gills. Every branch of these parapodia is internally supported by small spines, varying in number from two to four on the lower branch and from one to three on the upper one. Each branch also bears a tuft composed of numerous bristles.
Stratigraphic position and geographical distribution The genus is known in the Pennsylvanian (Upper Carboniferous, 300 million years ago) of Mazon Creek, Illinois. It occurs with a certain degree of frequency in these beds, especially in the species *E. zelus*, to which the example in the photograph also belongs.
Note Judging from the nature of its jaw apparatus, *Esconites* must surely have been a predator, but was itself the prey of primitive fish and sharks, as is illustrated by the remains of the jaws of *Esconites* found in the fossilized feces (coprolites) of vertebrates discovered in the same layer.

---

184 FOSSUNDECIMA

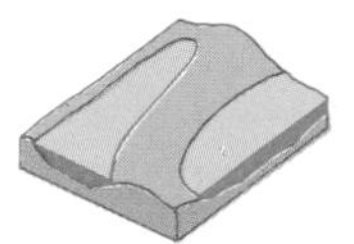

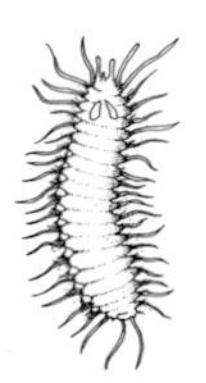

Classification Phylum Annelida, Class Polychaeta, Order Phyllodocida, Family Fossundecimidae.
Description A small annelid whose wide, segmented body has an average length of approximately 2.5 cm (1 in), with a maximum length of 4.5 cm (1¾ in). There are three short antennae on the head, as well as at least four long, tentaculiform cirri. One characteristic feature of this animal is the form of the jaws, which consist of a pair of strong, roughly triangular plates, with a serrated inner edge. The parapodia are biramous, with a short tuft of strong bristles on each spined branch.
Stratigraphic position and geographical distribution The genus occurs in the siderite nodules of the Mazon Creek beds in Illinois, which date from the Pennsylvanian (Upper Carboniferous, circa 300 million years ago). The particular species *F. konecniorum*, to which the example in the photograph belongs, is found in these beds.
Note *Fossundecima* was a predacious, probably omnivorous, polychaete; some examples have preserved within their stomach the remains of small invertebrates such as ostracod crustaceans, ingested by the animal in life.

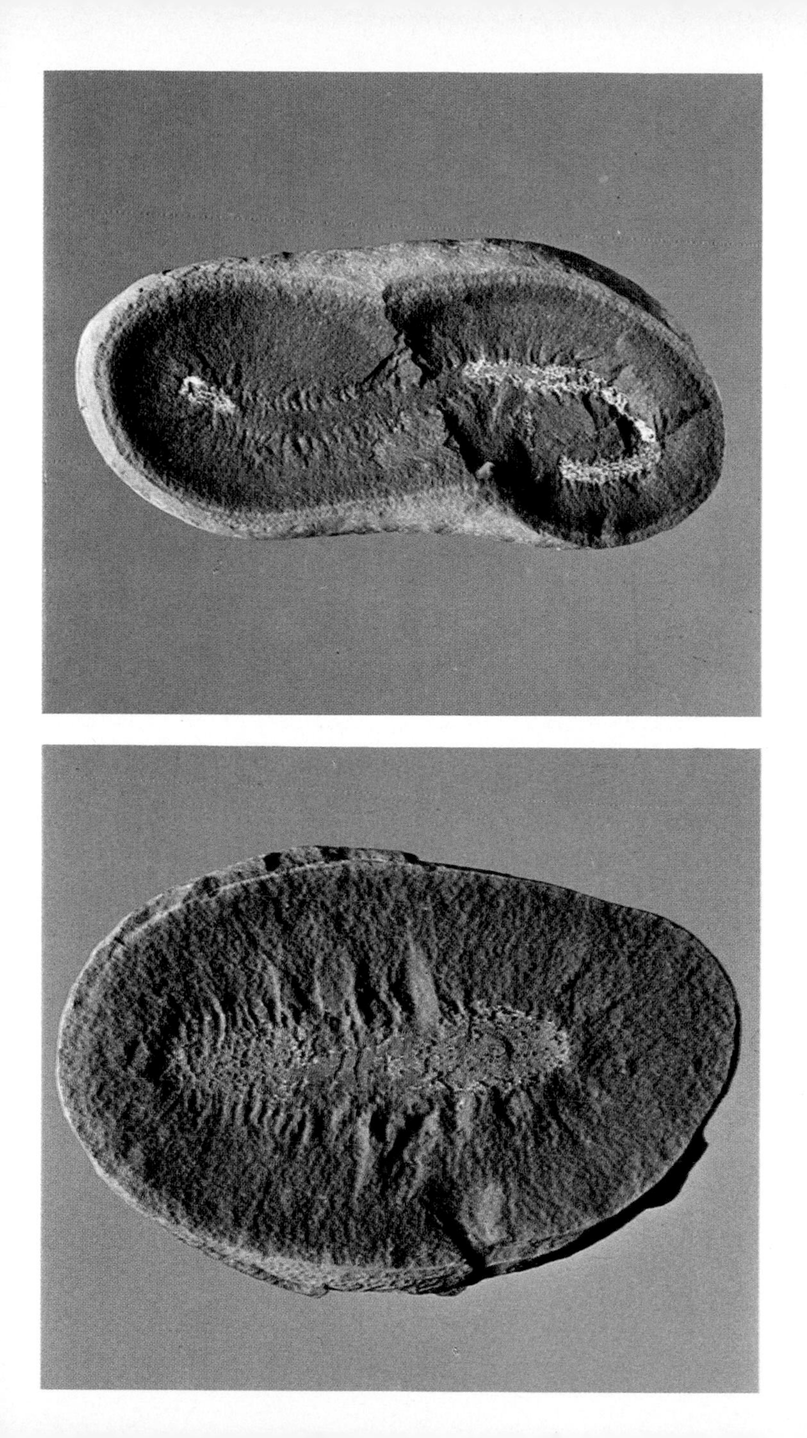

185 DICTYONEMA

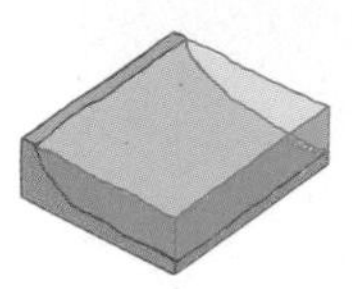

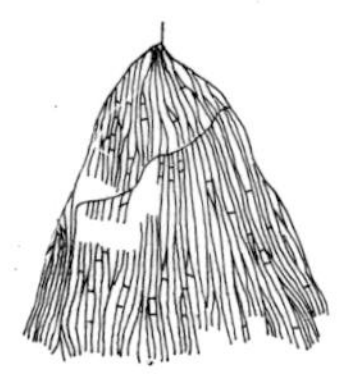

Classification Phylum Chordata, Class Graptolithina, Order Dendroidea, Family Dendrograptidae.
Description This graptolite, referred to the family Dendrograptidae, possesses rhabdosomes arranged like a fan; the stipes are regularly ramified. These ramifications are flexuous, arranged in a subparallel fashion and only rarely anastomose; they are joined together by transversally ranged dissepimentary structures. The autothecae carry spines.
Stratigraphic position and geographical distribution The genus *Dictyonema* first appears in the Upper Cambrian (510 million years ago), becoming extinct in the Lower Carboniferous (320 million years ago). It has a broad geographical distribution. The example in the photograph, which measures 8 cm (3¼ in), comes from Cambrian strata in Belgium.
Note The first representatives of these graptolites were equipped with a basal disk with which they affixed themselves to the substratum.

---

186 MONOGRAPTUS

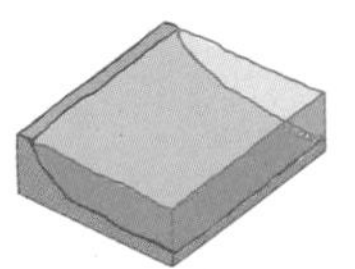

Classification Phylum Chordata, Class Graptolithina, Order Graptoloidea, Family Monograptidae.
Description The rhabdosome of *Monograptus* is uniserial. The thecae may be cylindrical, conical or in the shape of a flattened tube. Their position along the rhabdosome in the genus *Monograptus* is variable and represents a particular characteristic of the species: the thecae can, in fact, lie touching each other, sometimes fully overlapping; there are also several forms in which the thecae are situated quite far apart. The rhabdosome is usually straight, although it sometimes curves.
Stratigraphic position and geographical distribution The genus *Monograptus* is an excellent guide fossil for the Silurian (435–395 million years ago) and has been found only in Europe. The example in the photograph measures 2.5 cm (1 in) and comes from the Lower Silurian (435 million years ago) at Goni, Sardinia (Italy).
Note The family Monograptidae is represented by graptolites with thecae of various shapes and with a straight or curving rhabdosome. They are planktonic forms that occur fairly frequently in Italy, especially Sardinia.

187 DREPANASPIS

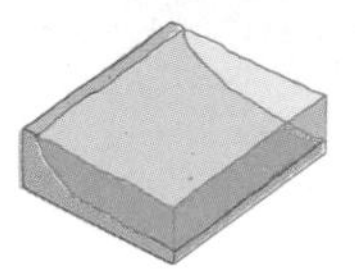

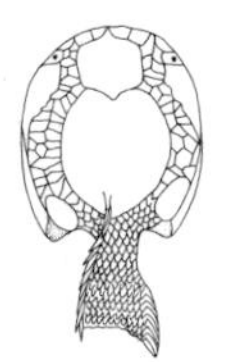

Classification Phylum Chordata, Class Agnatha, Order Heterostraci, Family Drepanaspidae.
Description A primitive representative of the Agnatha (jawless fish). Like many primitive agnathans of the order Heterostraci, *Drepanaspis* has a covering of plates to protect its body. The body is of flattened shape, much broader in the anterior region where the head and gills are situated; dorsally and ventrally this region has a broad, centrally positioned head shield, surrounded by a mosaic of smaller, polygonal platelets. The small eyes are anteriorly sited on each side, while the mouth is placed on the frontal extremity.
Stratigraphic position and geographical distribution The genus *Drepanaspis* is known from the European Lower Devonian (380 million years ago); complete examples, very well preserved, are found at Bundenbach in Germany. One such well-preserved example, in the photograph, measures approximately 30 cm (12 in) and belongs to the species *D. gemuendinensis*.
Note Heterostraci are an order of agnathans that lived from the Silurian to the Devonian. Isolated agnathan plates very similar to those of Heterostraci have been found in Upper Ordovician beds, which would make Heterostraci the oldest known vertebrates. Agnatha still survive in a few parasitic or semiparasitic forms, such as lampreys, which live attached to and sucking the blood of other fish; in the Lower Paleozoic, however, the agnathans were very widespread and diverse.

---

188 PTERICHTHYODES

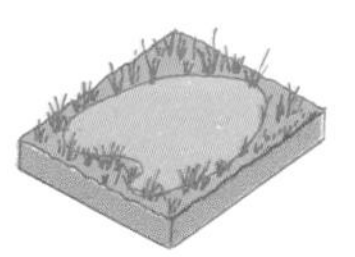

Classification Phylum Chordata, Class Placodermi, Order Antiarchi, Family Astrolepida.
Description Characteristic placoderm whose posterior body section is normally developed, covered in scales and with a heterocercal tail with a fleshy lobe ending upwards; the anterior part of the body, on the other hand, is much more unusual, being clad in a large shield of numerous plates covering the head and part of the trunk. The shield in *Pterichthyodes* is very tall in the thoracic region, over which it forms a pronounced hump; in the ventral region, however, it is flat. The head is short; the eyes and nostrils are placed close together on its upper part, while the mouth, situated below, is equipped with osseous, transversal plates.
Stratigraphic position and geographical distribution The genus is typical of the north European Middle Devonian (370 million years ago). The example in the photograph comes from the Scottish Devonian and measures approximately 20 cm (8 in).
Note The placoderms represent a class of fish that lived almost exclusively in the Devonian period, during which they were extremely widespread, especially in large inland expanses of fresh water. They were characterized by highly unusual forms whose monstrous appearance was caused by the shields in which they were covered. There were also large arthrodires, such as *Dunkleostus* (Class Placodermi), which had a shield measuring approximately 2 meters (6½ ft) out of a total body length of more than 4.5 meters (14½ ft).

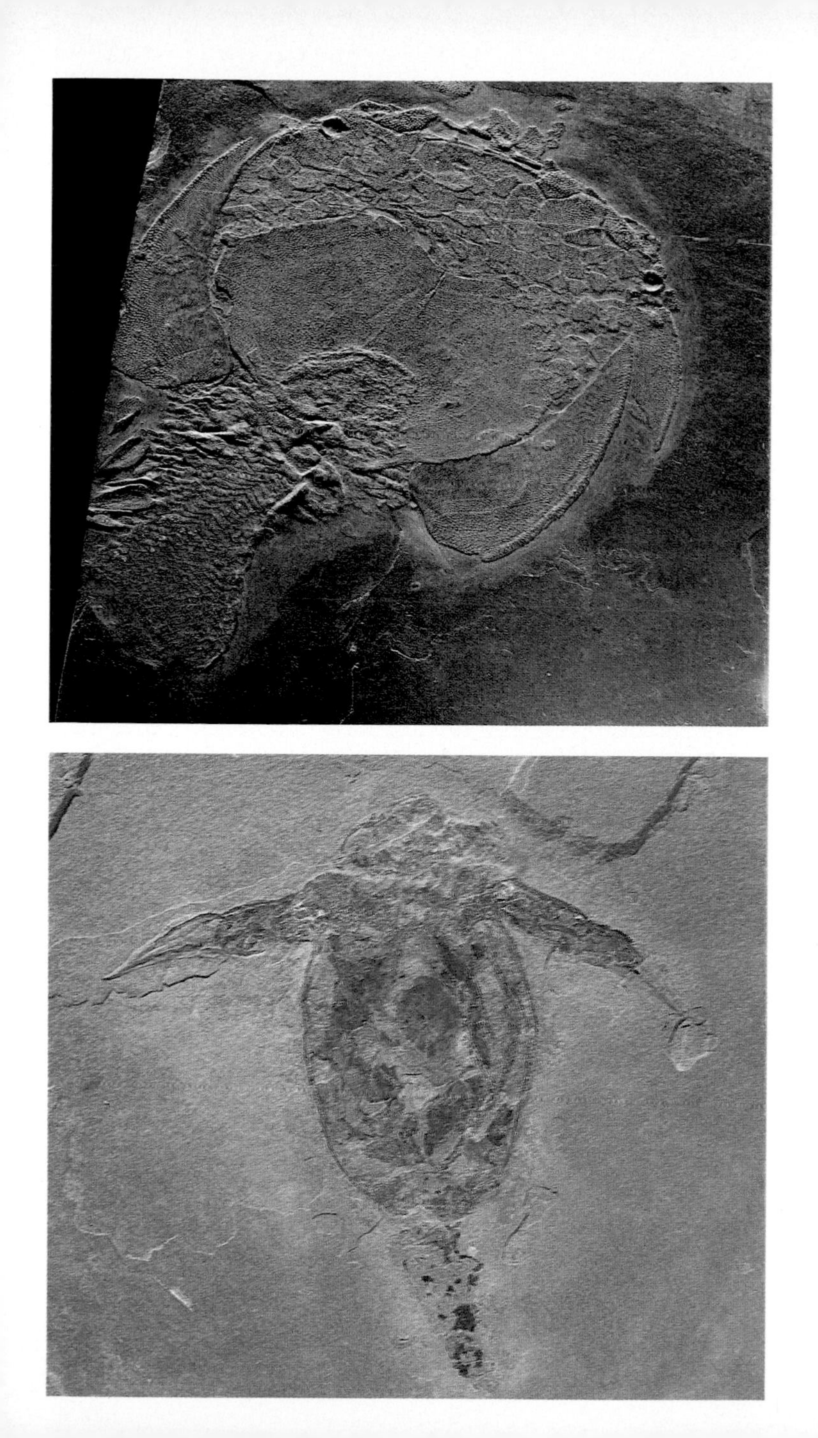

189 BOTHRIOLEPIS

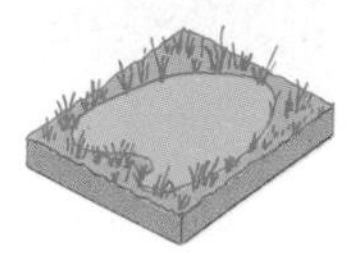

Classification Phylum Chordata, Class Placodermi, Order Antiarchi, Family Astrolepidae.
Description Typical form of placoderm and very similar in its general appearance to *Pterichthyodes*. *Bothriolepis* also has the anterior part of its body covered in a high shield, flattened in the ventral part, but without as pronounced a hump as in *Pterichthyodes*; it also differs in the posterior region of its body, which is naked and not covered in scales. At the sides of the shield there are two osseous appendages, projecting obliquely outwards, which articulate with the shield and are therefore mobile, contrary to the case of arthrodirian placoderms, which have fixed spines of this type. In some of these fish which have retained traces of their soft parts, the remains of lungs have been discovered.
Stratigraphic position and geographical distribution *Bothriolepis* lived in the Upper Devonian (360 million years ago) and was widespread throughout the world. The example in the photograph, *B. canadensis*, is approximately 8 cm (3¼ in) long and comes from the classic Lower Devonian sediments of Scaumenac Bay in Canada.
Note Placoderms such as *Bothriolepis* were freshwater fish, living on the bottom of large lakes. The development of their armor plating, like that of similarly equipped agnathans, has been related to their need to defend themselves from predators, in particular eurypterid arthropods, which must have been their principal enemies.

---

190 XENACANTHUS

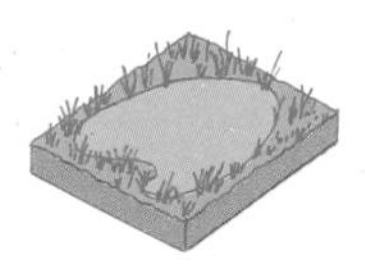

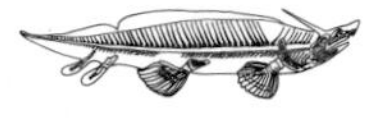

Classification Phylum Chordata, Class Chondrichthyes, Subclass Elasmobranchii, Order Pleuracanthodii, Family Xenacanthidae.
Description Typical representative of a primitive order of sharks which, unlike the majority of their fellows, lived in fresh water. The outline of this freshwater shark was long and slim, with a low fin that extended over a large part of its back. The paired fins were of a very primitive type, with an articulated, cartilaginous support, while the anal fin was typically formed of a double structure. One very characteristic feature of its head was the mobile spine, elongate, retroflex and equipped with a double row of denticles. The jaws possessed teeth with a central dome and two sharp, highly developed divergent points.
Stratigraphic position and geographical distribution The genus ranges from the Upper Devonian to the Upper Triassic (360–195 million years ago). It occurs in Europe, Australia and North and South America. The example in the photograph, which consists of teeth and elements of bone, comes from the Upper Carboniferous in Oklahoma and is attributed to the species *X. texensis*.
Note The genus, referred to by many as *Pleuracanthus*, forms part of an evolutionary line of sharks, the Pleuracanthodii, which, first appearing in the Devonian, were very abundant in the Carboniferous and Lower Permian and then went into a slow decline before disappearing completely at the end of the Triassic.

191 PSEUDODALATIAS

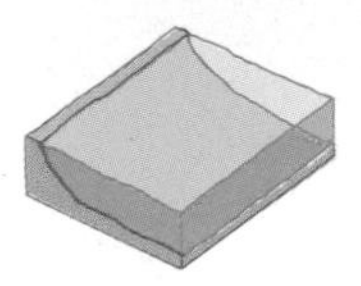

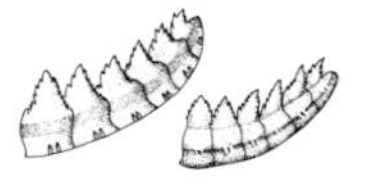

Classification Phylum Chordata, Subclass Elasmobranchii, Order Euselachii, Family Pseudodalatiidae.
Description A genus of fossil shark known solely from its teeth: these can be found singly or in small, complete sets of eleven teeth each. The sets are not very large in size, their maximum length being approximately 3 cm (1¼ in). A complete set comprises a central, or symphisial, tooth, with a laterally symmetrical outline, and five pairs of side teeth of asymmetrical outline. The individual teeth are laterally serrated, with four to six sharp serrations on each side.
Stratigraphic position and geographical distribution The genus is known from the single species *P. barnstonensis*, in the European Upper Triassic (200 million years ago), particularly in the bone beds of the English Rhaetian, where isolated teeth are found, and in the Norian-Rhaetian of the Alpine foothills in Lombardy (Italy), where complete sets occur, like the one in the photograph, which comes from Ponte Giurino, Bergamo (Italy).
Note This marine elasmobranch lived near the surface in inclosed tropical waters, whose depths were characterized by an extreme scarcity or even absence of oxygen.

---

192 ASTERACANTHUS

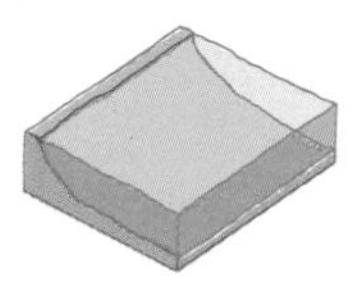

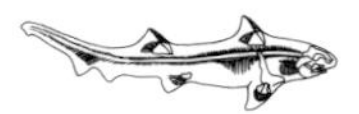

Classification Phylum Chordata, Class Chondrichthyes, Subclass Elasmobranchii, Order Selachii, Family Hybodontidae.
Description A unique type of shark because of its characteristic dentition, composed of low teeth whose cusps are sharp on the anterior teeth and more rounded on the posterior ones. Different from other sharks, whose teeth are arranged in rows inside the jaws, with only one row functioning at a time, in hybodonts such as *Asteracanthus* several sets were functional and able to cover extensive surfaces: the anterior ones were for tearing food, while the posterior ones were for grinding. These sharks also possessed robust spines on the dorsal fin. Spines and teeth represent the most common fossil remains.
Stratigraphic position and geographical distribution The genus ranges from the Upper Triassic to the Paleocene (200–55 million years ago). It has a broad geographical distribution, occurring in Europe, North Africa, Madagascar, the Middle East and North America. The spine in the photograph comes from the English Jurassic and belongs to the species *A. ornatissimus*; it measures approximately 14 cm (5½ in).
Note The group of hybodont sharks, which lived from the Carboniferous to the Paleocene, gave rise, according to many specialists, to the modern sharks. They were particularly widespread in seas of the Cretaceous period. Their dentition was adapted for grinding hard-shelled organisms such as Bivalvia and brachiopods, abundant in seas of the period.

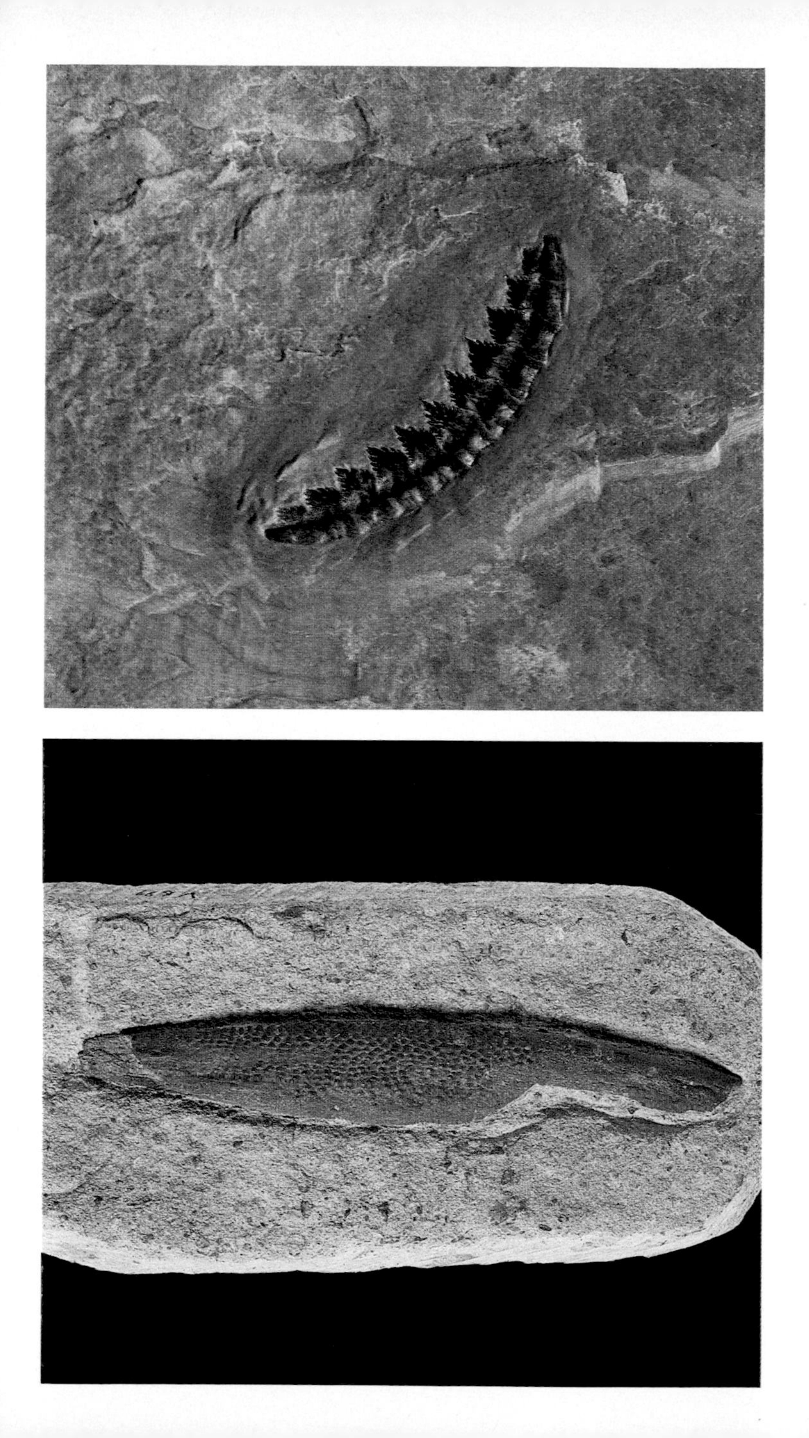

193 NOTIDANUS

Classification Phylum Chordata, Class Chondrichthyes, Subclass Elasmobranchii, Order Selachii, Family Notidanidae.

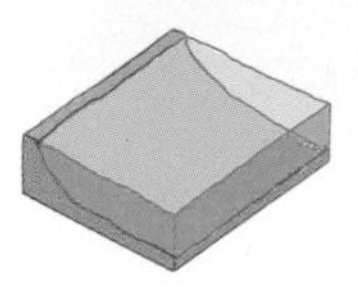

Description An ancient genus of shark that includes living members, characterized by the presence of six or seven gills, as opposed to the five found in other sharks. The living Notidanidae, such as *N. griseus*, are a long-bodied, six-gilled shark with a head as broad as the trunk; they are good swimmers and fearsome predators. Their teeth have sharp cusps and are markedly different in appearance from those of other sharks, being arranged like the teeth of a saw. They possess a single dorsal fin, positioned behind the ventral fins.

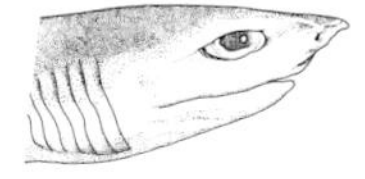

Stratigraphic position and geographical distribution The genus first appears in the Lower Jurassic (195 million years ago) and occurs in fossil form in a great number of horizons throughout the world: Europe, Asia, southwest Africa and North and South America. The photograph depicts teeth from the North American Cretaceous belonging to the species *N. anderssoni*.

Note Modern representatives of *Notidanus* show primitive characteristics such as the number of their gills, which is greater than in more highly evolved sharks. Unlike other primitive sharks such as the Pacific *Heterodontus*, which show typically hybodont features, *Notidanus* possess a single dorsal fin for the most part bereft of spines. Modern members of *Notidanus* are found in almost all the temperate to warm seas of the world.

---

194 CARCHARODON

Classification Phylum Chordata, Class Chondrichthyes, Subclass Elasmobranchii, Order Selachii, Family Isuridae.

Description The remains of this shark consist typically of pointed, triangular teeth, finely serrated at the cutting edges, which are found isolated in Tertiary sediments, sometimes in considerable numbers. The carcharodonts are large predators, like the present day white shark, which can exceed 8 meters (26 ft) in length. The body is fusiform, tapering in the posterior section, with a large head and mouth; the first dorsal fin, like the caudal one, is very developed. In the 17th century Nicholas Steno used the example of this shark to compare the fossilized teeth found in sediments with those of living forms, so demonstrating that fossils were not flukes of nature or spontaneous rock formations, but the remains of dead organisms that had been covered in sediment.

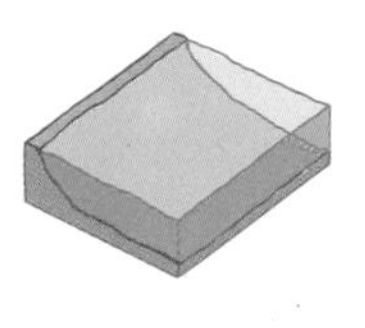

Stratigraphic position and geographical distribution The genus is first known in the Paleocene (65 million years ago), although there are doubtful remains dating from the Lower Cretaceous. It is widespread. The photograph shows a tooth of *C. megalodon* comes from the Maltese Upper Tertiary; it measures some 6 cm (2½ in).

Note The large size of the teeth of the fossil species *C. megalodon*, which can be up to 15 cm (6 in) long, led to the theory that there were sharks a full 30 meters (97½ ft) long living in Tertiary seas. Recent studies, however, have shown that *C. megalodon* must have been much smaller, though it could reach 15 meters (49 ft).

195 ODONTASPIS

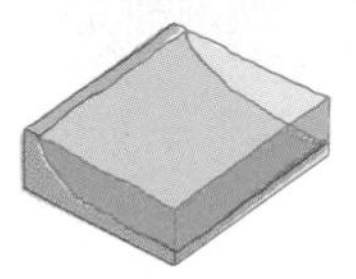

Classification Phylum Chordata, Class Chondrichthyes, Subclass Elasmobranchii, Order Selachii, Family Odontaspidae.

Description A genus of shark still represented in the modern world; its fossil remains comprise mainly teeth, long and thin in shape, which end in a sharp point and are often equipped with lateral cusps, also sharp, but much smaller in size. Many specialists include this genus within the genus *Carcharias*.

Stratigraphic position and geographical distribution The genus appeared during the Lower Cretaceous (140 million years ago) in Europe and soon spread throughout the world: by the end of the Cretaceous it had reached North and South America, Asia, Africa and New Zealand. The tooth in the photograph measures approximately 2.5 cm (1 in) and comes from Pliocene sediments in Florida.

Note The earliest known representatives of the Odontaspidae date from the Jurassic; the family reached its acme at the beginning of the Tertiary. They are now represented by such forms as *O.* (or *Carcharias*) *ferox* and *O. taurus. Odontaspiserox*, which is up to 4 meters (13 ft) long and weighs more than 300 kg (660 lb), possesses a streamlined body and lives in the eastern Atlantic and the Mediterranean, almost invariably in deep waters. It does, however, swim to the surface during the summer, becoming aggressive. *O. taurus* is fairly slender, with an elongate snout, and lives in the muddy depths of the Atlantic and the Mediterranean, where it searches for food in the form of molluscs, fish and crustaceans.

---

196 CYCLOBATIS

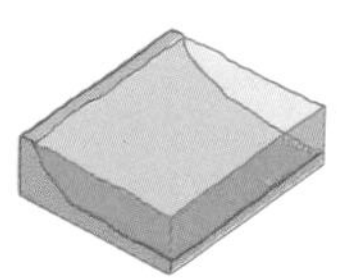

Classification Phylum Chordata, Class Chondrichthyes, Subclass Elasmobranchii, Order Rajiformes, Family Cyclobatidae.

Description A small rajiform. Rajiforms have a characteristically depressed body, flattened, with highly developed pectoral fins inserted into the sides of the trunk and the head and reaching as far as the snout; the head, the trunk and the pectoral fins thus form a single structure called the "disk." The disk of *Cyclobatis* has a characteristically circular shape, very large compared to the tail, which can remain quite small and is covered in two to three rows of pointed, triangular plates. The pectoral fins are supported by radial cartilages, the animal's most outstanding morphological feature, that may vary in number from 45 to 60. The pelvic girdle, the shoulder girdle and the cartilages of the long narrow cranium are often well preserved in fossils. The mouth has many small, flat teeth. The tail has no natatory dorsal fins.

Stratigraphic position and geographical distribution Different species of this genus exist in the Lebanese Upper Cretaceous (Cenomanian-Senonian, 100–65 million years ago); exceptionally well-preserved complete specimens have been found. The example, from the Lebanese Senonian at Sahel-Alma, measures about 9 cm (3½ in).

Note *Cyclobatis* was a benthonic organism living on the sea floor; it represents a specialized group of rays that is now completely extinct. The seas of the Lebanese Cretaceous also contained rays still represented today, such as *Rhinobatos*.

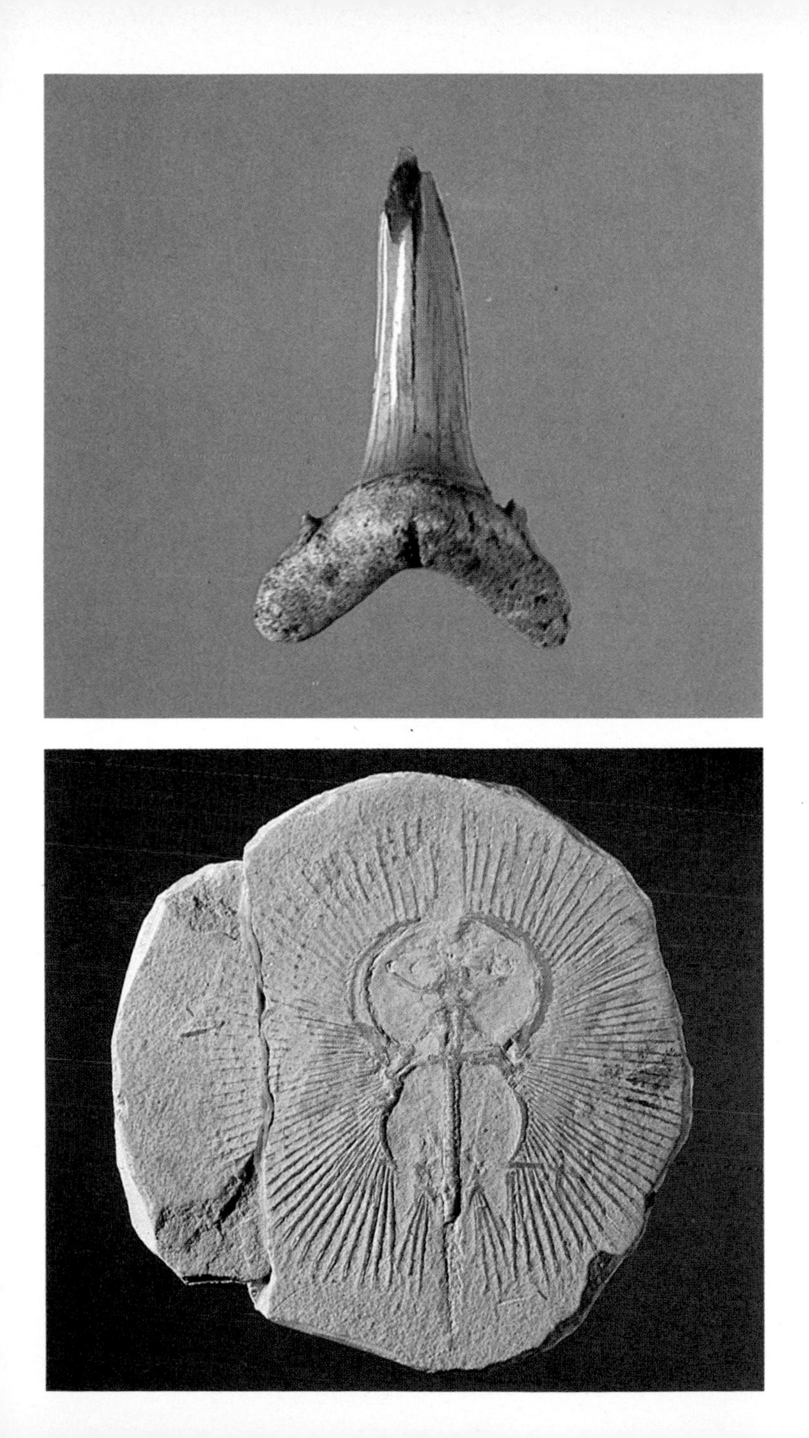

197 MYLIOBATIS

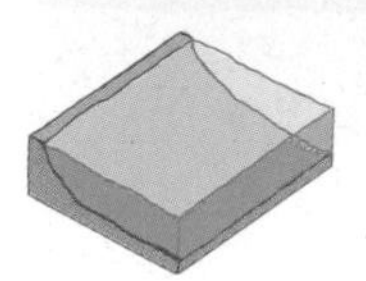

Classification Phylum Chordata, Class Chondrichthyes, Subclass Elasmobranchii, Order Rajiformes, Family Myliobatidae.
Description This genus of rays contains living forms, such as the so-called "eagle ray." The body is very flattened and the pectoral fins have a very broad lateral span that also embraces the wide, flat head. The long, whiplike caudal pedicle is highly developed. The teeth are the part of the body that survives most readily in fossil form, being more resistant than the cartilaginous skeleton: they are large and flat and arranged in serried, tongue-shaped rows creating a continuous surface suitable for crushing and grinding food. These teeth are more often found singly than in complete sets.
Stratigraphic position and geographical distribution The genus, which is still extant, is first known from the Upper Cretaceous (100 million years ago) and occurs in fossil form in Europe, North America, Africa, Asia and New Zealand. The photograph shows teeth of *Myliobatis*, measuring approximately 6 cm (2½ in), from Paleocene sediments in eastern Mali.
Note Modern Myliobatidae live mainly in warm seas, but they also occur in temperate seas such as the Mediterranean. The eagle ray lives close to sandy and muddy sea floors, feeding on molluscs, whose shells it is able to grind with its special dentition and strong musculature of the upper and lower jaw.

---

198 PHOLIDOPHORUS

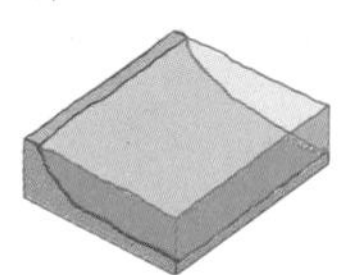

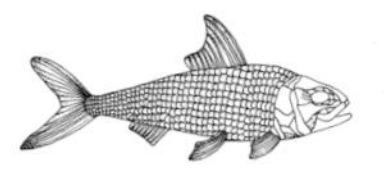

Classification Phylum Chordata, Class Chondrichthyes, Order Pholidophoriformes, Family Pholidophoriidae.
Description A holostean fish with an elongate, fusiform body and pointed snout; the dorsal fin is short, centrally situated on the back and more developed than the anal one. The pectoral and ventral fins are poorly developed. The body is covered in ganoid scales.
Stratigraphic position and geographical distribution The genus ranges from the Middle Triassic to the Upper Jurassic (220–140 million years ago). Examples of *Pholidophorus* are sometimes found in considerable numbers in many different horizons: the Middle Triassic in Germany and the Alpine foothills in Lombardy (Italy), and the Upper Triassic in England, the Tyrol and the Italian Alpine foothills; in the Lower Jurassic at Lyme Regis (England) and Württemberg (Germany) and in the Upper Jurassic in Bavaria, England and France; it is also found in South America (Argentina) and North and West Africa. The example in the photograph, which measures 6 cm (2½ in), comes from the Norian (Upper Triassic, 210 million years ago) in the Alpine foothills in Lombardy (Italy).
Note The genus belongs to an extinct group of holostean fish, which displayed characteristics very similar to those of the more evolved teleosts, the most widespread type of bony fish today, which supplanted the holosteans during the Jurassic. Holosteans are now represented by only two genera, *Amia* and *Lepidosteus*.

199 LEPTOLEPIS

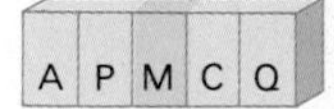

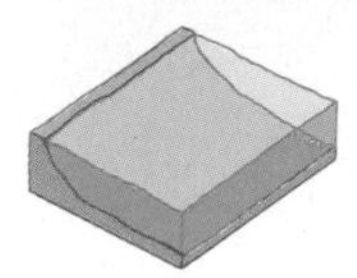

Classification Phylum Chordata, Class Osteichthyes, Order Lepidolepiformes, Family Leptolepidae.
Description A genus embracing Mesozoic forms that is in all probability heterogeneous. The body of this fish is normally small in size, with the overall appearance of a herring. It is elongate and tapering; the head is small, with a mouth equipped with small, sharp teeth. The dorsal fin is centrally situated on the back: it is not well developed, although it is larger than the caudal one. The pectoral and ventral fins are fairly well developed.
Stratigraphic position and geographical distribution The genus is typical of many Mesozoic horizons. It ranges from the Upper Triassic to the Upper Cretaceous (200–65 million years ago). It is found in the United States, South Africa, Madagascar and Asia, but it is especially typical of the European horizons, occurring in England (Somerset), France (Normandy, Burgundy, Lozère, Montpellier), Germany (Bavaria, Württemberg) and Spain. The example in the photograph comes from the Kimmeridgian at Solnhofen, Bavaria, and belongs to the species *L. knorrii*, which is typical of this layer; it measures approximately 12 cm (5 in).
Note The appearance of abundant remains of *Leptolepis* in Jurassic sediments, particularly those of the Upper Jurassic, signals the beginning of the fish world's domination by bony fish of the modern type.

---

200 PARANGUILLA

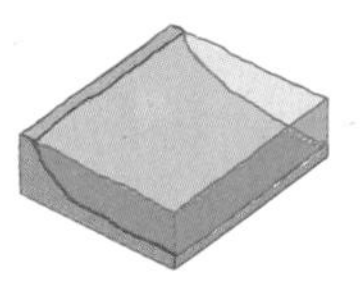

Classification Phylum Chordata, Class Osteichthyes, Order Anguilliformes, Family Xenocongridae.
Description A typical anguilliform, with an extremely elongate, almost snake-shaped body. The head is small, with a poorly developed gill slit. The dorsal fin is very long, reaching almost as far as the pointed tip of the tail; the anal fin is also well developed. The skin is completely devoid of scales. The teeth are small and sharp and must have been used for catching small animals.
Stratigraphic position and geographical distribution The genus is typical of the famous bed near Verona (Italy), dating from the Middle Eocene (Lutetian, 45 million years ago), where a variety of species of this genus has been discovered. The example in the photograph, approximately 45 cm (18 in) long, belongs to the species *P. tigrina*.
Note Numerous anguilliforms are found in the Bolca bed; these include some often very fine specimens of the genus *Paranguilla*, sometimes still possessing traces of their original color in speckles, as can be seen in the photograph.

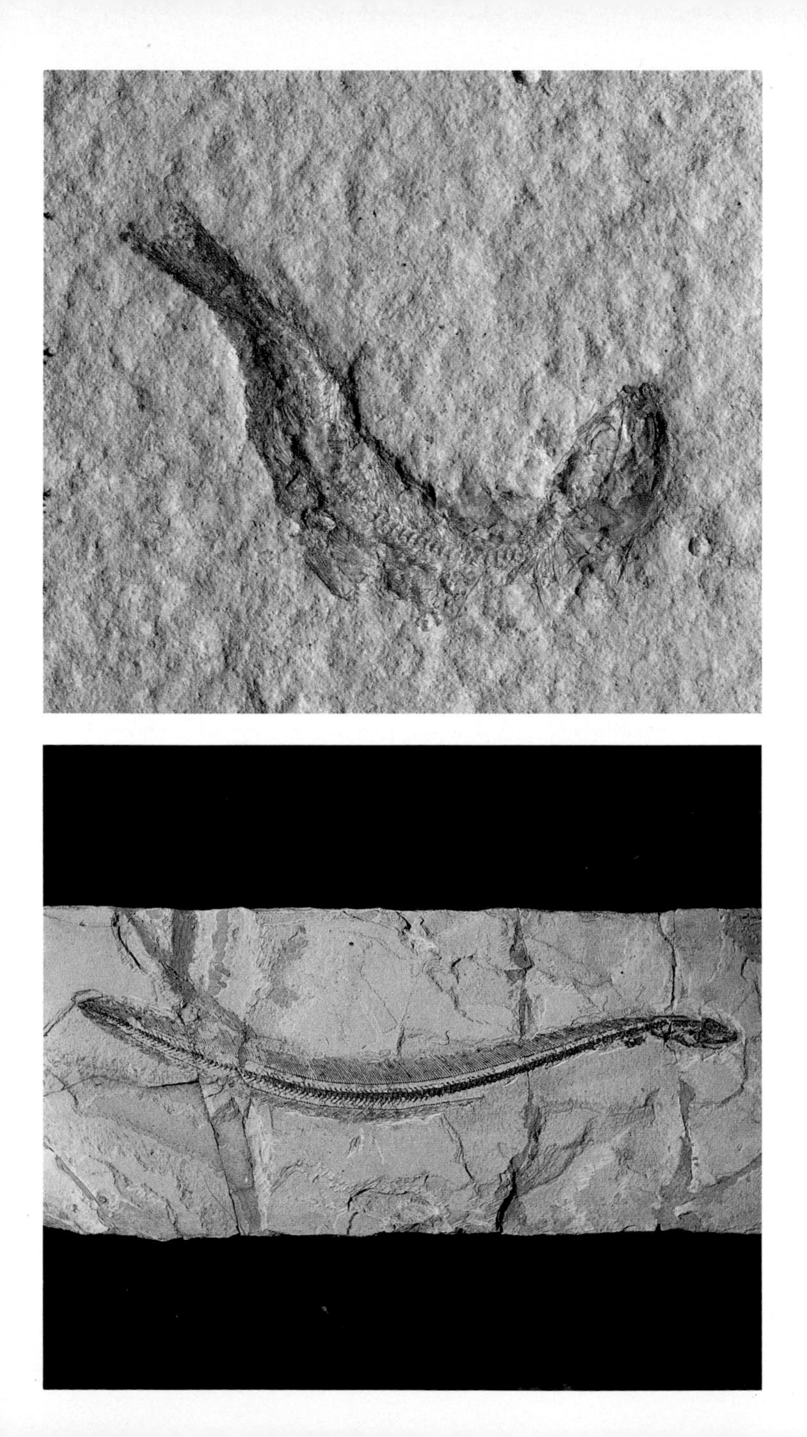

201 MENE

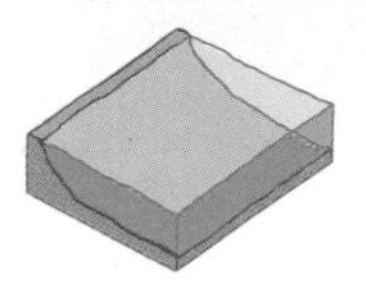

Classification Phylum Chordata, Class Osteichthyes, Order Perciformes, Family Menidae.
Description A characteristic tropical fish, with a laterally compressed body, rhomboidal in outline and generally lacking in scales; the scales, where present, are so small as to be invisible to the naked eye. The head has a pronounced and central bony crest shaped like a knife blade; the mouth is small, with small teeth that may disappear with age. The spinal column consists of 23 vertebrae. The dorsal and anal fins extend as far as the tail and are triangular in shape. The pectoral and pelvic natatory fins are placed very close together; two long, filiform rays emerge from the pelvic and natatory fins.
Stratigraphic position and geographical distribution The genus *Mene* is the sole representative of the family, known from the Eocene (50 million years ago) and still found in warm Asian seas. The example in the photograph, which belongs to the species *M. rhombea*, comes from the Eocene beds at Bolca (Verona) and is approximately 20 cm (8 in) long.
Note The famous Bolca fossil bed was originally a tropical lagoon whose waters teemed with life. Periodic mass mortality occurred among the organisms of this lagoon, and this has been attributed either to the effects of eruptions by nearby volcanoes or the sudden and excessive proliferation of toxic algae.

---

202 SPARNODUS

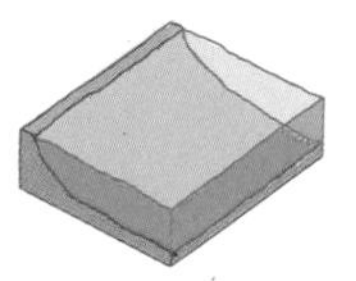

Classification Phylum Chordata, Class Osteichthyes, Order Perciformes, Family Sparidae.
Description A perciform of the family Sparidae, whose body displays the shape typical of this family; it is compressed at the sides and gibbous in the anterior portion. It is equipped with a single dorsal fin, well developed and supported by strong, spiny rays; the anal fin, by contrast, is half the length of the caudal, but similarly supported by spiny rays. The pectoral fins are long and well developed. The body is covered in large, finely wrinkled scales. The small mouth possesses a strong dentition formed of conical teeth. *Sparnodus* must have been a good swimmer and an active predator.
Stratigraphic position and geographical distribution The European genus *Sparnodus* ranges from the Middle Eocene to the Miocene (45–7 million years ago). Extremely fine and complete examples, belonging to a variety of species, occur in particular in the Bolca bed, Verona (Italy), which dates from the Middle Eocene. The example in the photograph, a member of the species *S. vulgaris* and measuring approximately 22 cm (8¾ in), also comes from the Bolca bed.
Note *Sparnodus* is an extinct genus of the family Sparidae, which today includes numerous forms living in warm and temperate seas throughout the world. Many of the family's representatives are well known for their gastronomic qualities: John Dory, dentex, sea bream and white bream.

203 OSTEOLEPIS

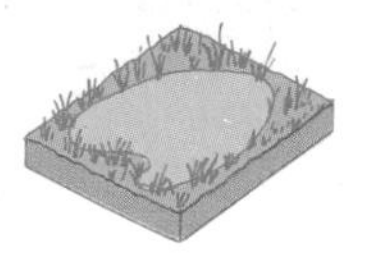

Classification Phylum Chordata, Class Osteichthyes, Subclass Sarcopterygii, Family Osteolepidae.
Description A typical Devonian crossopterygian fish with a tapering body densely covered in rhomboidal scales. The moderately stout head is broad and flattened; the tail is heterocercal and turns upward. There are two small dorsal fins; the pectoral and ventral fins are typically fleshy, with an internal skeletal support. The mouth possesses small, sharp teeth, which would have allowed the fish to prey on small invertebrates.
Stratigraphic position and geographical distribution The genus was widespread in the Middle Devonian in northern Asia and in the Middle and Upper Devonian in Europe (375–345 million years ago). The example in the photograph comes from Scottish Devonian sediments and measures some 12 cm (5 in).
Note The genus forms part of the group of rhipidistian Crossopterygii, freshwater fish that ultimately gave rise to the earliest terrestrial tetrapods, the amphibians, the first vertebrates endowed with proper ambulatory limbs. The rhipidistians became totally extinct at the end of the Paleozoic. Another group of Crossopterygii, the coelacanths, has survived up until the present day in a single species, the famous *Latimeria*, which lives solely in the Mozambique Channel off Africa.

---

204 CERATODUS

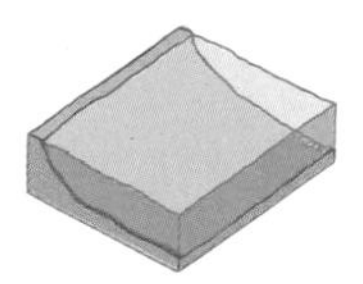

Classification Phylum Chordata, Class Osteichthyes, Subclass Sarcopterygii, Order Dipnoi, Family Ceratodontidae.
Description A Mesozoic dipnoan fish, whose most frequent fossil remains are its characteristic teeth, which are fused together to form flat plates and display varying degrees of serration on one side. The fossil genus is closely related to the living *Neoceratodus*, the Australian lungfish, which possesses a subcylindrical body covered in large, ossified scales. One characteristic feature of this animal is the fusion of the anal and dorsal fins with the caudal one; the paired fins, on the other hand, are sturdy and reminiscent of rudimentary limbs.
Stratigraphic position and geographical distribution The genus is widespread from the Triassic to the Paleocene (235–55 million years ago), although its period of greatest expansion occurred in the Triassic. It occurs in Europe, Asia, Australia, Africa and North and South America. The tooth in the photograph, which measures approximately 6 cm (2½ in), comes from Paleocene sediments in eastern Mali.
Note Dipnoi are fish equipped with a double system of respiration: they possess gills, as well as a lunglike organ. There are now only three surviving genera, living solely in fresh water in Australia, South America and Africa.

205 ERYOPS

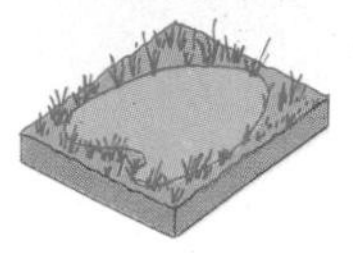

Classification Phylum Chordata, Class Amphibia, Order Temnospondyli, Family Eryopidae.
Description Of all the Rachitomi temnospondyls *Eryops* is certainly the best known, both because of the large number of examples discovered and also because of the many studies made of it. *Eryops* possessed a broad, flattened cranium, which reached 45 cm (18 in) in length and 34 cm (13½ in) in width and is so highly ossified that the sutures are hard to detect. The very large mouth was equipped with numerous teeth. The total length of the animal was around 2 meters (6.5 ft). Its very powerful limbs, together with its pelvic and scapular girdles, allowed it to walk without crawling.
Stratigraphic position and geographical distribution The genus *Eryops* has been found in Lower Permian strata (280 million years ago) in the states of Texas and New Mexico. The example in the photograph, which belongs to the species *E. megacephalus*, comes from the Lower Permian in Texas.
Note The genus *Eryops* represents the most advanced stage of adaptation to life on land among the temnospondyls.

---

206 BRANCHIOSAURUS

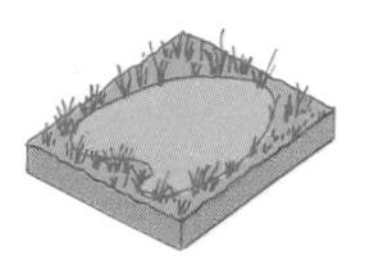

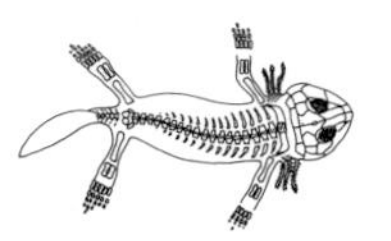

Classification Phylum Chordata, Class Amphibia, Order Temnospondyli, Family Eryopidae.
Description Members of the genus *Branchiosaurus* are relatively small animals (from 4–5 cm or 1.5–2 in to about 12 cm or 5 in). The very delicate skeleton is not greatly ossified. In some specimens where the fossilization process has been particularly successful, it is possible to make out, between the cranium and the scapular girdle, the leaflike external gills that the animal possessed in life. The cranium is of semicircular shape, flattened and equipped with two large eye sockets.
Stratigraphic position and geographical distribution The genus *Branchiosaurus* occurs in Upper Carboniferous sediments (290 million years ago) and became extinct in the Lower Permian (260 million years ago). It is limited in distribution to Europe. The example in the photograph, which belongs to the species *B. pterolei*, comes from the Carboniferous of Autun (France), and measures 3 cm (1¼ in).
Note The small skeletons of *Branchiosaurus* were for a long time the subject of much debate among amphibian specialists. They thought that its small size and the presence of external gills were the sort of characteristics that merited the inclusion of these amphibians in a group of their own. It has, however, proved possible to verify that *Branchiosaurus* is nothing more than the larval phase of large amphibians of the group *Eryops*.

207 MASTODONSAURUS

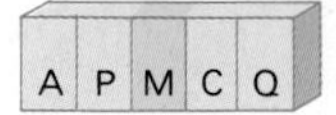

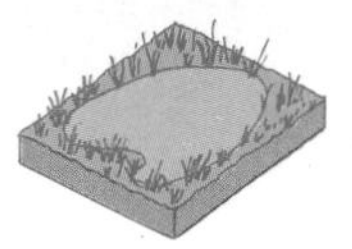

Classification Phylum Chordata, Class Amphibia, Order Temnospondyli, Family Capitosauridae.
Description *Mastodonsaurus* was an amphibian of truly exceptional size for the group to which it belongs: its cranium, for example, could be as much as 1.25 meters (4 ft) long and was posteriorly broad with a rather sharp anterior snout characterized by large, oval eye sockets. The shoulders of this animal were expanded in the distal region, which has led specialists to believe that *Mastodonsaurus*, despite its great bulk, was a good swimmer.
Stratigraphic position and geographical distribution The genus *Mastodonsaurus* has been found in Lower Triassic strata (230 million years ago) in Europe and occurs up to the Upper Triassic (200 million years ago), when it became extinct. Some specimens have been discovered in the African Upper Triassic. The cranium in the photograph, which belongs to the species *M. giganteus*, comes from the German Triassic.
Note With its 3-meter (9¾-ft) length, *Mastodonsaurus* was undoubtedly the largest amphibian ever to have lived on earth. This group of animals diversified into numerous forms, all of them large in size, during the Triassic.

---

208 SEYMOURIA

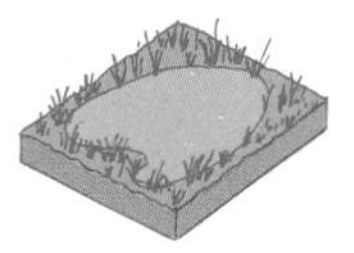

Classification Phylum Chordata, Class Amphibia, Order Anthracosauria, Family Seymouriidae.
Description *Seymouria* belongs to the group Anthracosauria and represents one of the most advanced stages in amphibian development; in fact, it possesses a mixture of amphibian characteristics, such as the area of the cranium in which the ear is situated, and reptilian ones, such as the presence of a single occipital condyle, vertebrae with the pleurocentra more developed than the intercentrum and limbs possessing the same number of phalanges as reptiles.
Stratigraphic position and geographical distribution The genus *Seymouria* occurs in Lower Permian sediments (280 million years ago) in the state of Texas. The example in the photograph comes from Texan beds and belongs to the species *S. baylorensis*.
Note *Seymouria* measured approximately 60 cm (23¾ in). It managed to keep its belly off the ground and its triangular cranium was equipped with numerous sharp teeth. Because of its mixture of characteristics, *Seymouria* has come to be regarded as marking the passage from amphibian to reptile. Its position within the class Amphibia has been the subject of lengthy debate, some paleontologists classifying it as a reptile.

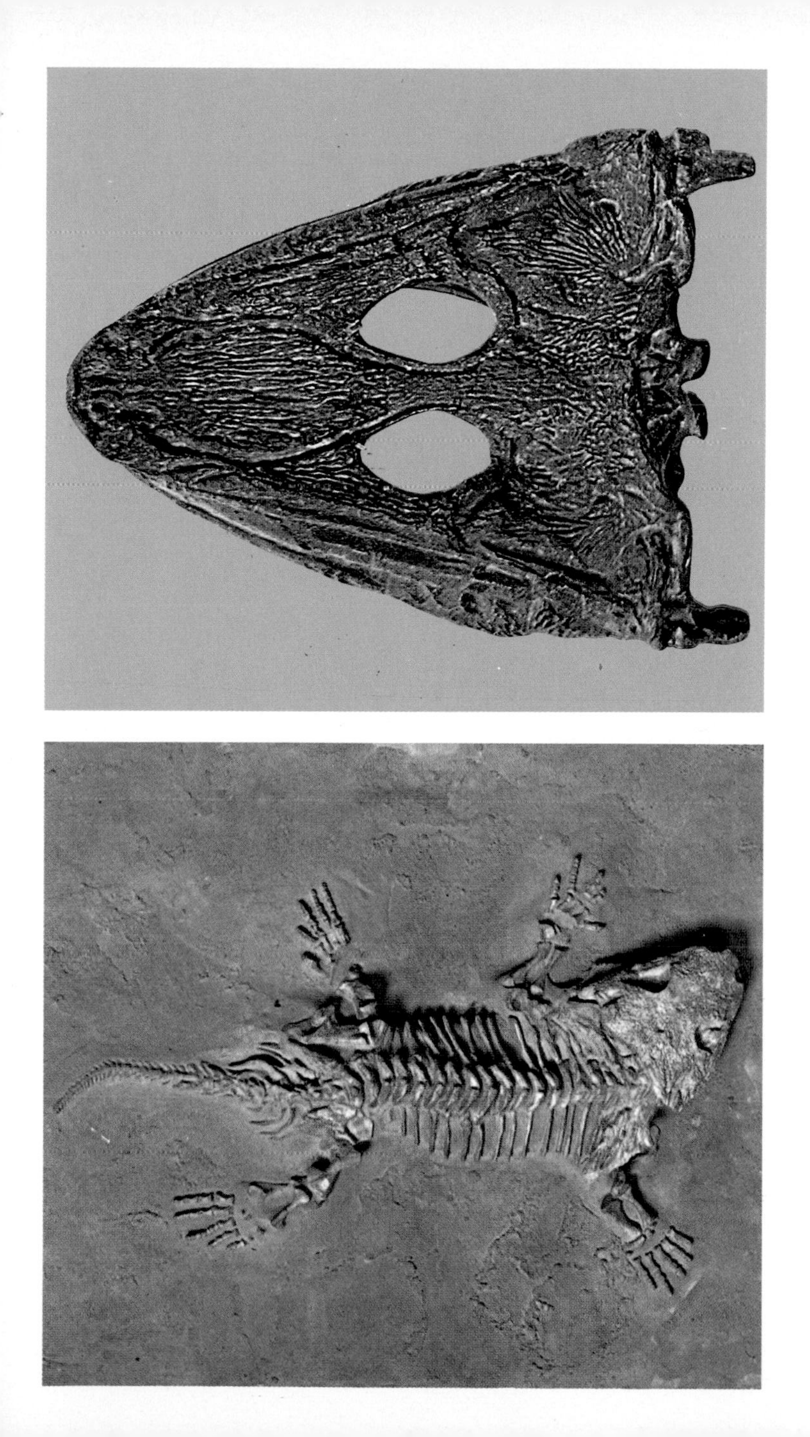

209 RANA

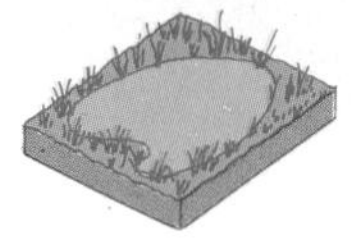

Classification Phylum Chordata, Class Amphibia, Order Anura, Family Ranidae.
Description The genus *Rana* contains small animals with a flattened, triangular cranium. The body is short and without a tail; the forelimbs are moderately developed, while the hind ones are very powerful, so as to allow the animal to jump. The phalanges of the hind limbs are also markedly longer than those present in the hand. Femurs and tibia are thin and elongate.
Stratigraphic position and geographical distribution The genus *Rana* first appears in Europe in the Eocene (55 million years ago), but does not occur in Asian, African or North American sediments until the Miocene (22.5 million years ago). It is still living today. The example in the photograph belongs to the species *R. pueyoi* and comes from the Miocene at Libros in Spain; it measures 15 cm (6 in).
Note Anurans are known from the Upper Carboniferous (290 million years ago) in forms that do not diverge greatly from their modern counterparts, even though their hind limbs are not yet adapted to jumping. They are excellent paleoecological indicators, being closely linked to lacustrine life.

---

210 LABIDOSAURUS

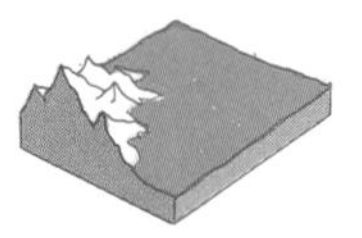

Classification Phylum Chordata, Class Reptilia, Order Cotylosauria, Family Captorhinidae.
Description From the top of its head to the tip of its tail, this captorhinid reptile reached a length of 70 cm (27½ in). Its compressed head, triangular in shape, was characterized by a dentition composed of numerous pointed, triangular-shaped teeth. The cranium had no aperture and the orbicular fossa was characterized by a bony structure that was used to protect the eyes or to focus on images. The humeri and femurs were massive and projected horizontally.
Stratigraphic position and geographical distribution The genus *Labidosaurus* has been found in Lower Permian outcrops (280 million years ago) in the state of Texas. The example in the photograph, which belongs to the species *L. hamatus*, comes from the Texan strata.
Note The earliest captorhinid was discovered in Upper Carboniferous rocks (280 million years ago) in Europe and is assigned to the genus *Solenodonsaurus*.

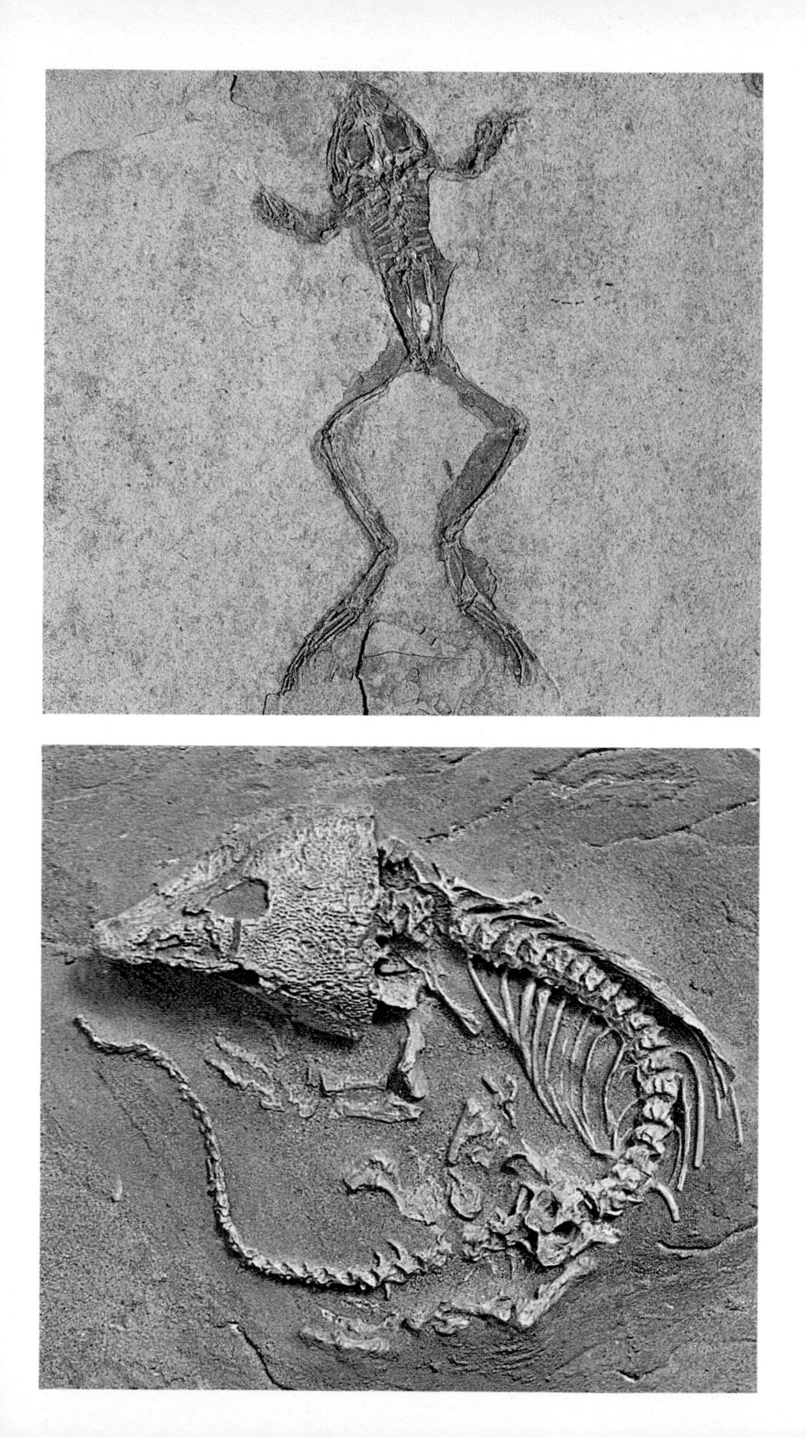

211 PAREIASAURUS

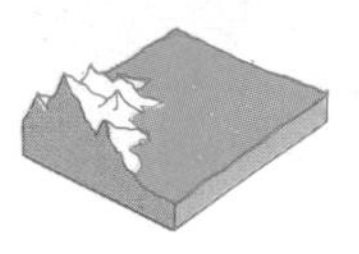

Classification Phylum Chordata, Class Reptilia, Order Cotylosauria, Family Pareiasauridae.
Description A reptile of extremely archaic and massive build, reaching approximately 3 meters (9¾ ft) in length; it possessed a cranium adorned with numerous bony protuberances that were particularly prominent on the lower mandibles. The dentition was very simple, being composed of a series of identical, peglike teeth. The overall appearance of the cranium was massive and grotesque, with no aperture. The limbs are similarly massive, with broad feet and hands that served to increase the animal's stability during movement.
Stratigraphic position and geographical distribution The genus *Pareiasaurus* has been found in Permian strata (260 million years ago) in the South African Karroo. The example in the photograph, a member of the species *P. baini*, comes from these African strata.
Note The cotylosaurs comprise very primitive forms of reptile that first appear in the Carboniferous, with some surviving up until the Triassic. Very extensive remains of these animals have been found in South Africa. During the Upper Permian they reached Europe.

---

212 MESOSAURUS

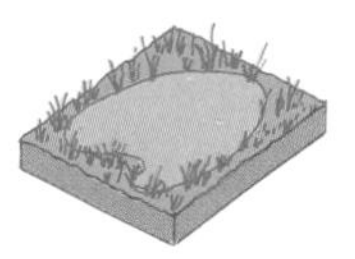

Classification Phylum Chordata, Class Reptilia, Order Mesosauria, Family Mesosauridae.
Description A small reptile, some 40 cm (16 in) long, characterized by a cranium with large numbers of regularly spaced teeth, arranged in such a way that an upper tooth fits into the gap between adjacent lower teeth and vice versa. The long tail was used by the animal while swimming, whereas the limbs are unmodified for an aquatic life.
Stratigraphic position and geographical distribution The genus *Mesosaurus* is found in Lower Permian sediments (280 million years ago) in South America and southern Africa. The example in the photograph, of the species *M. tumidus*, comes from Brazil and measures 20 cm (8 in).
Note *Mesosaurus* lived in the lakes that occurred on the South American and African continents during the Lower Permian, when the two land masses were joined together to form a single continent. These fossils are part of the evidence that validates the plate tectonics theory.

213 TRIONYX

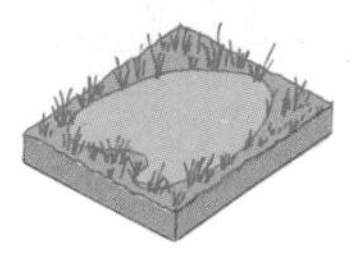

Classification Phylum Chordata, Class Reptilia, Order Chelonia, Family Trionychidae.
Description A reptile of small or medium size, whose body is encased in armor with an anterior aperture to allow the head to emerge, and a posterior one to let the tail protrude. The osseous part of the carapace is created by the expansion of eight vertebrae and as many ribs surrounded by marginal plates. The head is retractable and the pelvis is independent of the sturdy shell and of the plastron that protects the venter.
Stratigraphic position and geographical distribution The genus *Trionyx* appeared in the Upper Jurassic (160 million years ago) in Europe; in the Lower Cretaceous (140 million years ago) it reached Asia and North America; during the Miocene it reached Africa, and in the course of the Pliocene (5 million years ago) it disappeared from the continent of Europe and appeared in India. The example in the photograph, which belongs to the species *T. capellinii*, comes from the Oligocene of Monteviale near Vicenza (Italy); it measures 32 cm (12¾ in).
Note The order Chelonia, to which *Trionyx* belongs, consists of reptiles with no cranial fenestrations. The first chelonians appeared in the Triassic in the form of *Triassochelys* and chelonians are still widespread in both marine and terrestrial environments today.

---

214 DREPANOSAURUS

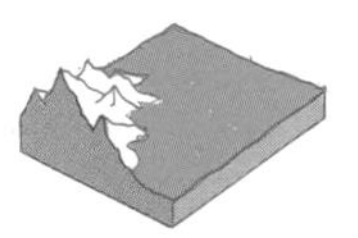

Classification Phylum Chordata, Class Reptilia, Subclass Lepidosauria, Order Squamata.
Description Small lepidosaur reptile that reached a length of some 50 cm (19¾ in). The nature of this animal's cranium is unknown, since all the specimens so far discovered are lacking this part of the body, which has perhaps been the object of depredation in all cases. The tail was powerful and highly developed; it possessed a peculiar hook at its tip. Both the forelimbs and hind limbs were highly developed and characterized by claws; one of these, present on the ungual phalange of the second digit of the forelimb, was extremely well developed and sickle-shaped.
Stratigraphic position and geographical distribution The genus *Drepanosaurus* is known from remains discovered in outcrops of Zorzino limestone, a Norian limestone (210 million years old), at Endenna in Val Brembana (Bergamo, Italy), which is also the source of the example in the photograph.
Note *Drepanosaurus* was a terrestrial reptile and probably able to move by running on its hind legs. It was undoubtedly very active, but its diet is unknown due to the absence of any cranium.

215 ASKEPTOSAURUS

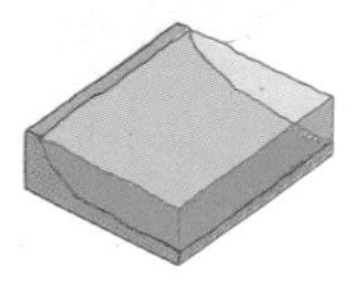

Classification Phylum Chordata, Class Reptilia, Order Eosuchia, Family Thalattosauridae.
Description A reptile similar to a crocodile that reached 2 meters (6½ ft) in length, with a long, slim body well adapted to the aquatic life. It possessed an elongate cranium characterized by the presence of numerous pointed teeth and by nostrils placed far back, near the eyes. The body extended into a long, mobile tail, which must have been used for swimming, given the reduced dimensions of the fore- and hind-limbs when compared with those of the tail. Its legs would certain not have allowed the animal to move quickly over dry land.
Stratigraphic position and geographical distribution The genus *Askeptosaurus* is known from the Middle Triassic (220 million years ago) at Monte San Giorgio in the canton of Ticino (Switzerland). The example in the photograph, which belongs to the species *A. italicus* and measures 2 meters (6½ ft) in length, comes from the ichthyolithic schist formation at Besano, near Varese (Italy).
Note This animal's adaptation to the aquatic life was not complete, and so *Askeptosaurus*, like the marine iguana, spent part of its time on land, breeding and laying its eggs. It fed on fish and small reptiles.

---

216 PTERODACTYLUS

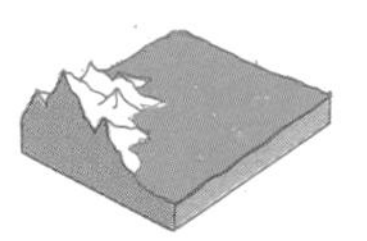

Classification Phylum Chordata, Class Reptilia, Order Pterosauria, Family Pterodactylidae.
Description Small flying reptile, characterized by an elongate cranium with reduced dentition and supported on a long neck. The combined length of the cranium and neck was almost twice that of the animal's body, which had no tail. The forelimb bore just four digits, the fourth highly developed and supporting the thin membrane used for flight.
Stratigraphic position and geographical distribution The genus *Pterodactylus* is first known from the Upper Jurassic (150 million years ago) of Europe, Africa and Asia. The latest discoveries in Europe and Asia have been made in strata dating from the Lower Cretaceous. The example in the photograph, which belongs to the species *P. antiquus*, comes from the famous Solnhofen beds in Germany and measures approximately 10 cm (4 in).
Note Pterodactyls, and probably all flying reptiles, were animals incapable of the same sort of flight as birds; the membrane borne by the fourth digit acted only as an aid to gliding. In order to take flight they went up onto cliffs and launched themselves into the air like gliders. As far as can be deduced, this membranous structure was probably very delicate. Flying reptiles became extinct at the end of the Cretaceous.

217 COELOPHYSIS

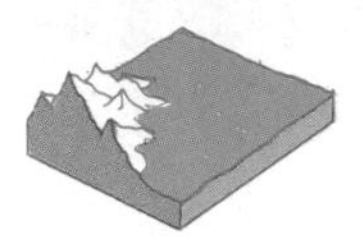

Classification Phylum Chordata, Class Reptilia, Order Saurischia, Family Procompsognathidae.
Description This compsognath was a small saurischian dinosaur: it measured only 2.6 meters (8½ ft) and had long, slender and very sturdy rear legs, a feature typical of running bipedal animals. The forelimbs were small, by contrast, and equipped with claws used for catching hold of prey. The cranium was narrow and elongate, with developed jaws containing numerous sharp teeth. The cranium was fairly light in weight, and was supported by a highly mobile neck. The tail was long and flexible, while the pelvis was of saurischian type. *Coelophysis* probably laid small eggs. One of the strangest anatomical features of this animal was the presence of certain hollow bones, similar to those found in birds.
Stratigraphic position and geographical distribution The genus *Coelophysis* has been discovered in Upper Triassic strata (200 million years ago) in the U.S. The photograph shows a Triassic rock from the state of New Mexico with numerous skeletons belonging to the sole known species, *C. bauri*.
Note *Coelophysis* was a very agile predator, which moved swiftly by either walking or running on its back legs. The tail was held pointing straight out behind to counterbalance the weight of the cranium.

---

218 ALLOSAURUS

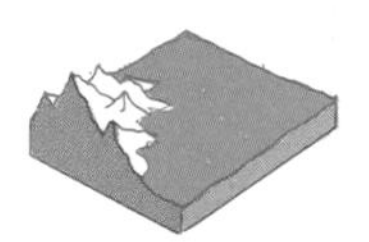

Classification Phylum Chordata, Class Reptilia, Order Saurischia, Family Megalosauridae.
Description A bipedal, carnivorous dinosaur that reached a length of seven meters (23 ft); in biped position its cranium was approximately 2.5 meters (6½ ft) off the ground. It possessed a highly developed cranium, lightened by numerous fenestrae and containing numerous very sharp, conical teeth. The forelimbs were reduced in size and equipped with three robust claws, while the hind limbs were highly developed and bore thin, powerful digits that demonstrate the degree of agility with which this animal moved. There was also a long tail serving to balance the weight of the cranium and the body itself while running.
Stratigraphic position and geographical distribution The genus *Allosaurus* has been discovered in Upper Jurassic sediments (160 million years ago) in the U.S. The example in the photograph, which belongs to the species *A. fragilis*, comes from the state of Montana. *Allosaurus* has also been found in the states of Utah, Wyoming, Arizona, New Mexico, Nevada and Colorado.
Note The bones of this animal are black because, during fossilization, salts of manganese became deposited in the cavities, thereby mineralizing them.

219 TYRANNOSAURUS

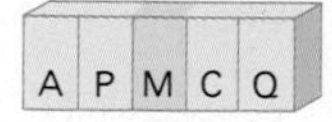

Classification Phylum Chordata, Class Reptilia, Order Saurischia, Family Tyrannosauridae.
Description A carnivorous dinosaur of large dimensions, roughly 15 meters (48.75 ft) long, reaching a height of 6 meters (19.5 ft) and weighing around 10 tons. An inhabitant of highlands and plains, it successfully hunted such herbivorous dinosaurs as, for example, ceratopsians and "duck-billed dinosaurs." The general structure of the body is that of a predator: bipedal standing posture, highly developed hind limbs and very reduced forelimbs equipped with claws. The cranium, which was highly developed and lightened by numerous fenestrae, possessed a dentition of very strong conical teeth. These were constantly renewed: there were always fully developed teeth as well as growing ones so that teeth remained sharp for tearing.
Stratigraphic position and geographical distribution The genus *Tyrannosaurus* has been discovered in Upper Cretaceous sediments (75 million years ago) in North America, and footprints of the animal have also been found in Asia. The photograph shows a reconstruction of the cranium of *Tyrannosaurus*.
Note As its front legs were virtually useless, *Tyrannosaurus* probably used its hind legs to disembowel its victims after having chased them and then killed them with its powerful jaws.

---

220 PLATEOSAURUS

Classification Phylum Chordata, Class Reptilia, Order Saurischia, Family Plateosauridae.
Description A large carnivorous dinosaur that reached 8 meters (26 ft) long and 5.5 meters (18 ft) high in a biped position. The cranium was very elongate and contained large numbers of simple, spatulate teeth. The front legs were less developed than the hind ones, but the animal was not completely bipedal. The cranium, relatively small in size, was borne on a long neck that gave *Plateosaurus* a rather comical appearance. The hind legs, like the front ones, possessed five digits, even though on the hind legs the first and fifth digits were smaller than the others.
Stratigraphic position and geographical distribution Abundant discoveries of the genus *Plateosaurus* have been made in Triassic rocks, some 200 million years old, in Germany, the source of the example in the photograph. Other discoveries, also in Triassic rocks, have been made in France and southern Africa.
Note During the Jurassic, the plateosaur group gave rise to the sauropods, with forms such as *Brontosaurus* and *Diplodocus*.

221 BRONTOSAURUS

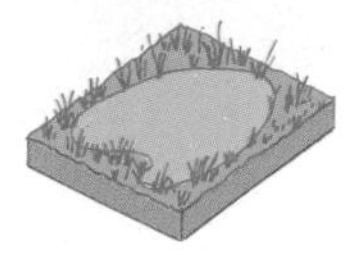

Classification Phylum Chordata, Class Reptilia, Order Saurischia, Family Titanosauridae.
Description A large saurischian, sauropod dinosaur, reaching as much as 18 meters (58½ ft) in length. The overall shape of the body was fairly squat. It possessed a tiny cranium, situated at the end of a long neck. The cranium contained a dentition composed of simple, peg-shaped teeth in the first two-thirds of the jaws. *Brontosaurus* was a herbivore.
Stratigraphic position and geographical distribution The genus *Brontosaurus* has been found in the Morris Formation in the U.S., dated to the Upper Jurassic, approximately 160 million years old. The photograph shows a *Brontosaurus* skeleton from the state of Montana.
Note *Brontosaurus* was a herbivore with a very small cranium in comparison with its body size; it fed on aquatic plants and it is thought that it must have eaten continuously in order to maintain the great bulk of its body. The titanosaurs represent the only truly giant form of dinosaur.

---

222 DIPLODOCUS

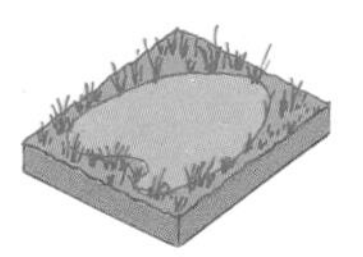

Classification Phylum Chordata, Class Reptilia, Order Saurischia, Family Titanosauridae.
Description A gigantic, herbivorous dinosaur, which attained lengths of up to 25 meters (82 ft). Its body had a much slimmer outline than *Brontosaurus*, with a much longer neck and tail. The cranium was small and contained peg-shaped teeth only in the anterior section of the mouth; these teeth were probably used to graze the aquatic plants on which it fed. The legs were much longer than those of other titanosaurs and fairly slender overall. The vertebrae of the neck were slightly elongate.
Stratigraphic position and geographical distribution The genus *Diplodocus* has been found in Upper Jurassic sediments (160 million years ago) in the state of Montana, the origin of the skeletal remains shown in the photograph.
Note Some paleontologists maintain that sauropods, to compensate for the fact that they were so heavy, lived in the ponds and marshes that abounded in North America during the Jurassic.

223 PSITTACOSAURUS

Classification Phylum Chordata, Class Reptilia, Order Ornithiscia, Family Ceratopsidae.

Description *Psittacosaurus* was a very small ornithiscian reptile of the family Ceratopsidae, no more than 1 meter (39 in) long. It had a biped stance and possessed a low, short cranium, anteriorly modified into a structure similar to that of a parrot's beak. It possessed forelimbs slightly less developed than the hind ones, which bore the whole weight of the body. The teeth were situated towards the back of the oral cavity and arranged in sets.

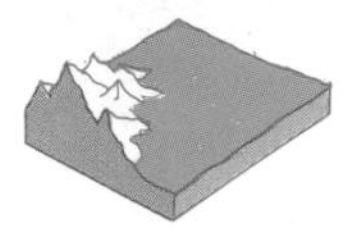

Stratigraphic position and geographical distribution The genus *Psittacosaurus* is known from remains unearthed in strata dating from the Upper Cretaceous (80 million years ago) in Mongolia. The *Psittacosaurus* skull shown in the photograph, which measures approximately 15 cm (6 in), is also from this area.

Note This small protoceratopsian may have been in the group that gave rise to the large ceratopsian or "horned" dinosaurs such as *Triceratops*.

---

224 KENTRUROSAURUS

Classification Phylum Chordata, Class Reptilia, Order Stegosauria, Family Stegosauridae.

Description An ornithiscian dinosaur of medium size. It had a quadriped stance and its body was very strong and compact. The cranium was small, with reduced cavities and peg-shaped teeth of equal shape and size. The diet of this dinosaur was herbivorous. Its body had a characteristic series of spines that acted as protection against attacks by its carnivorous contemporaries. The hind limbs were longer and more developed than the forelimbs, but the animal nevertheless walked exclusively on all fours. The tail was equipped with sharp spines that undoubtedly acted as a very efficient means of self-defense.

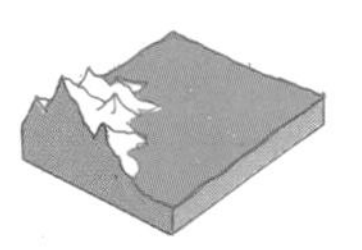

Stratigraphic position and geographical distribution The genus *Kentrurosaurus* has been found in beds of the Upper Jurassic, approximately 160 million years old, in East Africa. The skeleton in the photograph comes from Tanzania.

Note Stegosaurs are a group of ornithiscian dinosaurs characterized by the presence of protective spines, plates and tubercles on their bodies. The most famous is *Stegosaurus*, which inhabited the North American continent during the Jurassic.

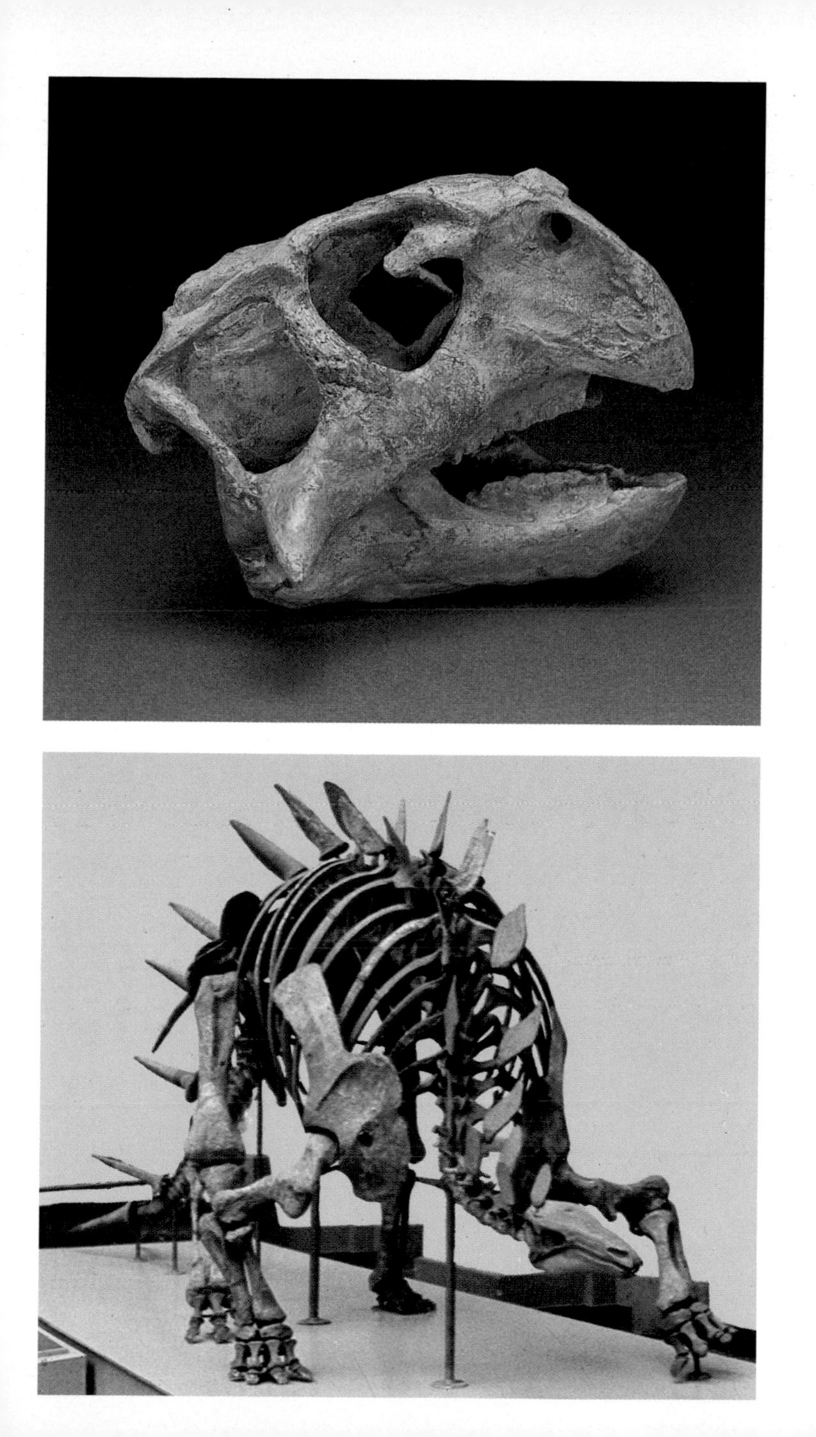

225 TRICERATOPS

Classification Phylum Chordata, Class Reptilia, Order Ornithiscia, Family Ceratopsidae.
Description A large, herbivorous dinosaur, characterized by a massive cranium equipped with a pair of horns placed over the eye sockets, a further horn on its nose and a "beak." The cranium is posteriorly accompanied by a bony frill: this acted as a means both of protecting the neck from attacks by predators and also of retaining the powerful neck muscles that supported the cranium. The arrangement of the forelimbs was typically reptilian, while the hind limbs were almost columnar. The spinal column was rigid, and there was a powerful tail that may also have played a defensive role.
Stratigraphic position and geographical distribution The genus *Triceratops* is found in Upper Cretaceous strata (70 million years ago) in North America. The species *T. prorsus*, which appears in the photograph, comes from North American strata and measures 8 meters (26 ft) in length.
Note *Triceratops* inhabited the grassy plateau that once stretched across the continent of North America. It fed on vegetation, which it gathered using its powerful "beak," while its horns, as well as being a means of defense, could also be used for stripping leaves off trees.

---

226 CERESIOSAURUS

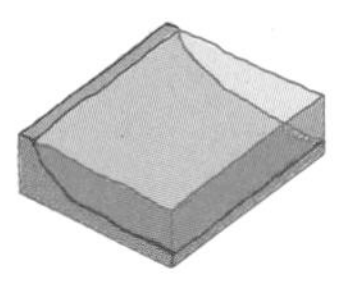

Classification Phylum Chordata, Class Reptilia, Order Saurischia, Family Nothosauridae.
Description A large nothosaur reptile that could reach up to 3.8 meters (12 ft) in length, characterized by a triangular cranium with sharp, conical teeth. The cranium was supported by a long flexible neck. The body was streamlined and equipped with a long powerful tail. The hind limbs and forelimbs were only slightly modified for aquatic life because this reptile spent part of its life on dry land. It fed on smaller reptiles that it found on the beach and on fish and cephalopods that it caught in the sea.
Stratigraphic position and geographical distribution The genus *Ceresiosaurus* has been discovered in Italian and Swiss Middle Triassic (220 million years ago) strata at Besano (Val Ceresio, near Varese in Italy) and at Monte San Giorgio (Canton of Ticino in Switzerland). The species *C. calcagnii*, shown in the photograph, was found at Monte San Giorgio.
Note The skeleton in the photograph is complete, with only the cranium affected by small cracks; the fossil was accompanied by a certain number of *Pachypleurosaurus*, which had been preserved at the same time. It is clear from its physiology that it was more agile in water than on dry land.

227 PACHYPLEUROSAURUS

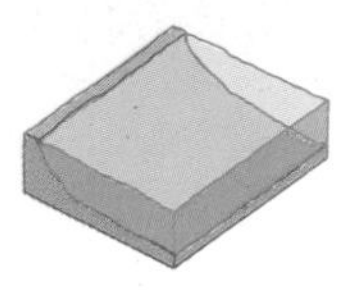

Classification Phylum Chordata, Class Reptilia, Order Saurischia, Family Pachypleurosauridae.
Description A small nothosaur reptile that existed in only two known forms, not yet differentiated at species level. The large-headed form reached a length of 40 cm (16 in), and the form characterized by a tapering cranium reached 1 meter (39 in). It was an active carnivore that spent much of its time in water; it is possible that it possessed webbed digits. Its limbs, however, showed very little adaptive modification for marine life, which meant that the animal moved easily on dry land as well. Its long tail was undoubtedly used during swimming.
Stratigraphic position and geographical distribution The genus *Pachypleurosaurus* is found in outcrops of Italian and Swiss Middle Triassic (220 million years ago) at Val Besano (Val Ceresio, near Varese in Italy) and at Monte San Giorgio (Canton of Ticino in Switzerland). The photograph shows two examples of the species *P. edwardsii* from the ichthyolithic shales at Besano.
Note *Pachypleurosaurus* is a typical fossil of the Besano shales, which at one stage were industrially exploited. Paleontologists believe that the young of this genus lived in groups, which is why they are found close to each other, whereas the adults were solitary individuals. The young hatched from eggs on dry land.

---

228 CRYPTOCLEIDUS

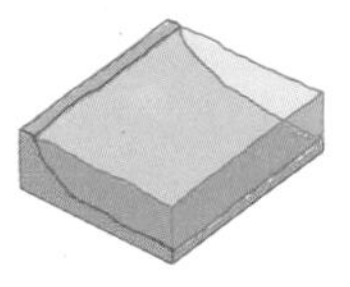

Classification Phylum Chordata, Class Reptilia, Order Sauropterygia, Family Plesiosauridae.
Description A large marine reptile that reached approximately 3.5 meters (11½ ft) in length. Its body was elongate and streamlined. The cranium, equipped with large numbers of sharp teeth, was of modest size and was borne on a mobile neck. Both the forelimbs and hind limbs were modified to become paddles for use in swimming; the humeri and the femurs were reduced in length and expanded at the distal extremities, as were the ulna, radius, tibia and fibula. The hand and the foot were extremely modified; there was an increase in the number of phalanges making up the hand and foot (a condition termed "hyperphalangia"), thereby turning the whole limb into a paddle for swimming. The tail was moderately developed and also used for swimming.
Stratigraphic position and geographical distribution The genus *Cryptocleidus* is characteristic of the European Upper Jurassic (160 million years ago). The example in the photograph, which belongs to the species *C. oxoniensis*, comes from the English Upper Jurassic.
Note Plesiosaurs were active carnivores that caught their prey by sudden movements of their mobile necks rather than by swimming, for they were relatively slow animals. They were totally adapted to the aquatic life.

229 PSEPHODERMA

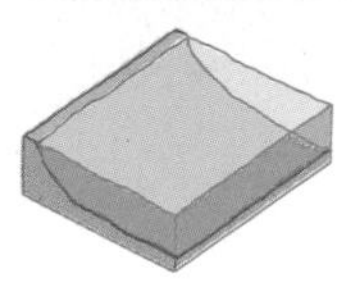

Classification Phylum Chordata, Class Reptilia, Order Placodontia, Family Placochelyidae.
Description A marine reptile adapted to the aquatic life, whose body was similar to a turtle's. It possessed a triangular cranium equipped with kidney-shaped teeth situated on the palate. It had a carapace formed of large numbers of bony plates firmly fused together; this carapace was marked by three longitudinal ridges, one in the center and one on each side. A tail protruded from the posterior end of the carapace.
Stratigraphic position and geographical distribution The genus *Psephoderma* is known from Upper Triassic strata (200 million years ago) in Europe. The example in the photograph belongs to the species *P. alpinum* and comes from the Norian at Endenna (Bergamo in Italy); it is approximately 1.2 meters (about 4 ft) long.
Note The placodonts are a group of reptiles that lived in the seas of the Upper Triassic. Numerous forms are known, some of them equipped with a sturdy shell and others with their bodies exposed. A common characteristic of these genera is the presence of palatine teeth, either round or kidney-shaped, used to crack the shells of the molluscs that they extracted from the sea floor by means of their pointed or shovel-shaped "beak;" in animals with shovel-shaped beaks the snout also bore teeth anteriorly. The placodonts became extinct, leaving no descendants.

---

230 ICHTHYOSAURUS

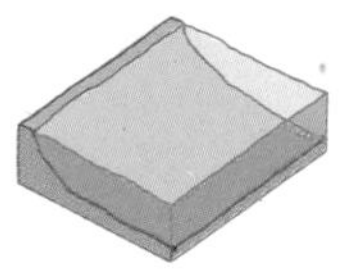

Classification Phylum Chordata, Class Reptilia, Order Ichthyosauria, Family Ichthyosauridae.
Description A marine reptile as hydrodynamic as a fish, characterized by an elongate cranium terminating in a long, sharp beak and equipped with conical, striated teeth that rest not in alveoli but within a furrow in the maxillary and mandibular bones. The cranium has a very small fenestration. The spinal column curves downwards in the caudal region, modified by the structure of the tail, which moved horizontally. The forelimbs, like the hind ones, are modified into true swimming flippers. The animal never left the water and gave birth to live young.
Stratigraphic position and geographical distribution The genus is known through discoveries made in Toarcian sediments (Lower Jurassic, 180 million years ago) at Holzmaden (Germany), the origin of the example in the photograph.
Note There is a famous fossil in the Holzmaden layer of a female carrying young before birth. From the fossils that occur so abundantly in this German fossil bed paleontologists have obtained a physical profile of *Ichthyosaurus*, since the impressions of the soft parts of animals have been preserved in this unique deposit. The Jurassic ichthyosaurs were derived from the Triassic mixosaurs, which are found at Besano (Varese in Italy) and Monte San Giorgio (Canton of Ticino in Switzerland).

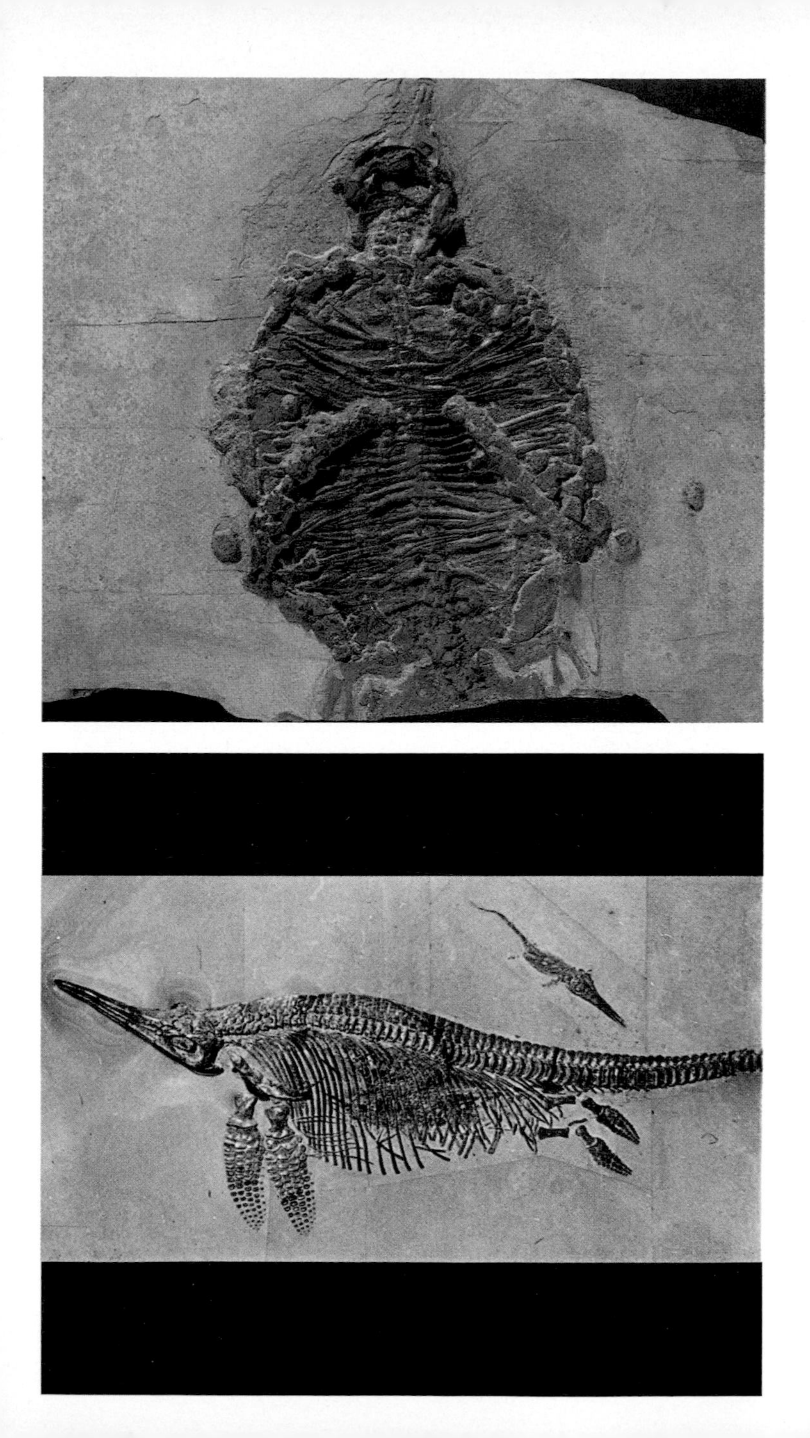

231 EDAPHOSAURUS

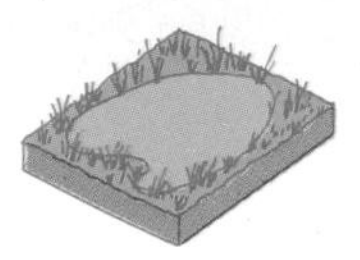

Classification Phylum Chordata, Class Reptilia, Subclass Synapsida, Order Pelycosauria, Family Edaphosauridae.
Description A medium-sized herbivorous reptile, characterized by a very small cranium equipped with numerous small, sharp teeth. The body is highly developed; in the thoracic region there are long spines that are extensions of the vertebrae. These vertical spines have much smaller lateral spines arranged at right angles to the spinal process, from which they emerge almost regularly. The limbs are robust and the feet highly developed. There is a long tail, which is also vertically developed to form a continuous crest.
Stratigraphic position and geographical distribution The genus *Edaphosaurus* has been discovered in Lower Carboniferous rocks (290 million years ago) in Europe and North America. The example in the photograph comes from the Permian in the state of Texas and measures approximately 3 meters (9¾ ft).
Note *Edaphosaurus* was prey to other mammal-like reptiles, among them *Dimetrodon*. It was a highly specialized herbivore, probably feeding on lacustrine plants.

---

232 COTYLORHYNCHUS

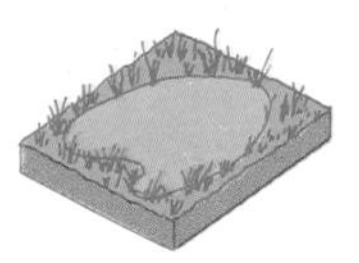

Classification Phylum Chordata, Class Reptilia, Subclass Synapsida, Suborder Edaphosauria, Family Caseidae.
Description *Cotylorhynchus* was a synapsid reptile belonging to the suborder Edaphosauria. It was a large, herbivorous animal, reaching 3 to 4 meters (9¾–13 ft) in length. It possessed a small cranium with a few blunt teeth suited to grinding vegetable matter. It must have been a very clumsy animal. Its head was tiny in comparison to its body, which included, among other things, an extremely long tail. The feet of this mammal-like reptile were highly developed, as were the forelimbs and the hind limbs, whose task it was to support the great weight of the body.
Stratigraphic position and geographical distribution The genus *Cotylorhynchus* has been found in the San Angelo Formation (Lower Permian, 280 million years old) in North America; it disappeared soon afterwards, in the Middle Permian. The example in the photograph comes from the San Angelo Formation.
Note *Cotylorhynchus* was one of the most primitive pelycosaurs and also one of the largest. Its diet was probably aquatic plants.

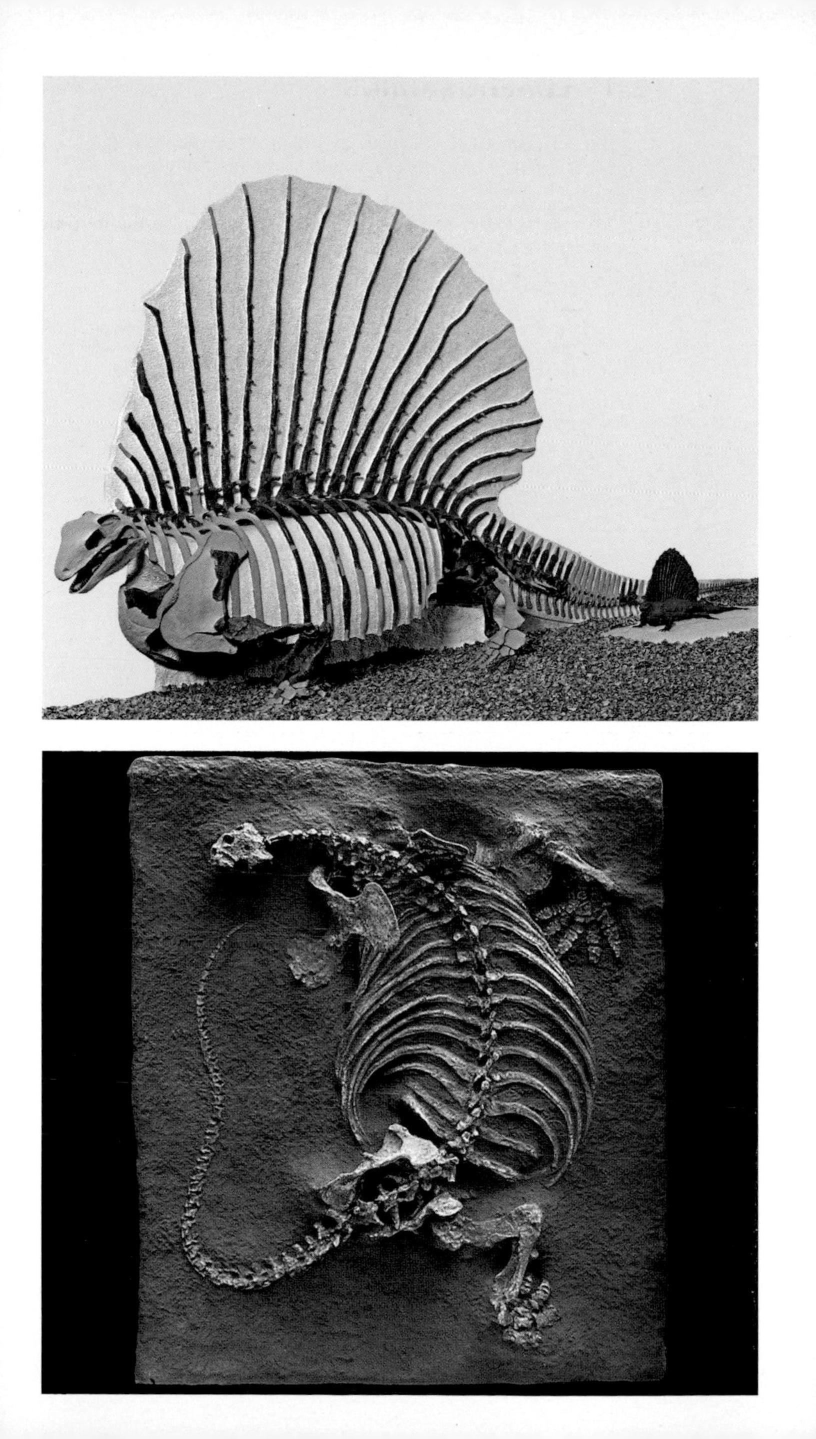

233 CYNOGNATHUS

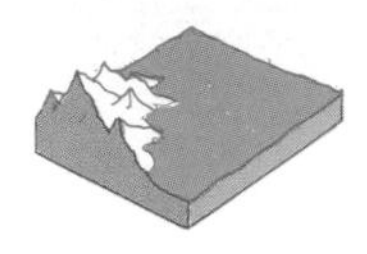

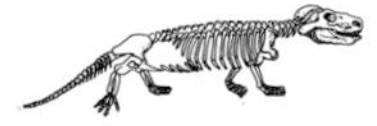

Classification Phylum Chordata, Class Reptilia, Subclass Synapsida, Order Therapsida, Family Cynognathidae.
Description *Cynognathus*, a carnivorous form of mammal-like reptile, is one of the most evolved of these strange animals. The cranium had numerous triangular teeth, suitable for tearing the flesh of the animal's victims, and long, sharp canines. Its auditory sense was highly developed. The general structure of the body was extremely muscular, agile and quick; running was its main means of catching prey.
Stratigraphic position and geographical distribution The genus *Cynognathus* (the photograph shows a detail of the cranium) has been found in Lower Triassic strata (230 million years ago) in South Africa. Its size was more or less equivalent to that of a dog.
Note Mammals are derived, via a series of gradual modifications, from these mammal-like reptiles. *Cynognathus* is the one that is anatomically closest to the mammals.

---

234 ARCHAEOPTERYX

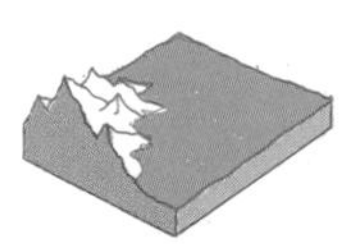

Classification Phylum Chordata, Class Aves, Order Archaeopterygiformes, Family Archaeopterygidae.
Description A bird characterized by the presence of an elongate, triangular cranium with teeth fixed in alveoli (as in reptiles). It possessed a long tail equipped with feathers on both sides. The sternum of this particular bird was not carinate, a sign of a poor aptitude for flight, while its forelimbs possessed three digits equipped with claws. Among its reptilian characteristics is the absence of bones containing air pockets; the presence of feathers is an avian characteristic.
Stratigraphic position and geographical distribution The genus *Archaeopteryx* is known through discoveries made in Upper Jurassic sediments (150 million years ago) at Solnhofen, Germany, the origin of the example in the photograph (*A. lithographica*).
Note *Archaeopteryx* represents the link between birds and reptiles. This early bird may not have flown. Its feathered limbs may have been used for warmth and catching prey. It must have been very clumsy, more like a guinea fowl (which it resembles in size, among other things) than a bird of flight.

235 DIATRYMA

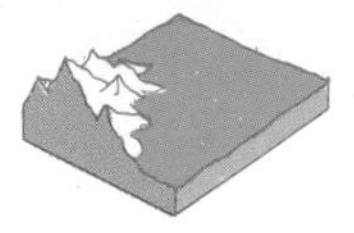

Classification Phylum Chordata, Class Aves, Order Diatrymiformes, Family Diatrymatidae.
Description A bird, more than 2 meters (6½ ft) high with a robust, highly developed cranium bereft of teeth, but equipped with a strong beak. The hind limbs were massive, the forelimbs very small. The body was covered not in feathers, but in black, downy filaments (which have been discovered in fossil state).
Stratigraphic position and geographical distribution The genus *Diatryma* is first known from the Upper Paleocene (60 million years ago) and became extinct during the Middle Eocene (45 million years ago). Its geographical spread is restricted to North America and Europe. The photograph shows *D. gigantea* (to the left), together with two other flightless birds: *Struthio camelus* (center) and *Dinormis maximus* (right).
Note *Diatryma* was incapable of flight, and had become adapted to terrestrial life. During the course of the Paleocene and Eocene it was the uncontested predator of small mammals. During that period the only rivals of *Diatryma* were crocodiles.

---

236 ALCA

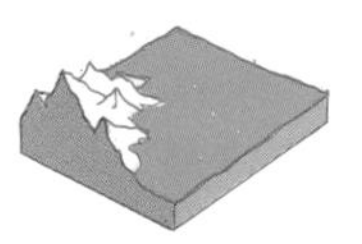

Classification Phylum Chordata, Class Aves, Order Charadriiformes, Family Alcidae.
Description A bird adapted to marine life, a good swimmer but totally incapable of flight. Its body was similar in general appearance to that of a penguin, although *Alca* is in no way related to the penguin. It grew to a size of approximately 50 cm (20 in). The cranium had a highly developed beak and large eye sockets situated far to the back. The sternum was large and very broad. The small forelimbs were used during swimming whereas the hind limbs were long and sturdy.
Stratigraphic position and geographical distribution The genus *Alca* first appears in the Miocene (20 million years ago). The example in the photograph (a detail showing the cranium) comes from the Icelandic Quaternary and belongs to the species *A. impennis*.
Note This bird was driven to extinction by man, having been widely hunted for its flesh, its fat and its soft feathers. The last individual was killed on June 3, 1844 at Eldey, Iceland.

GIANT FLIGHTLESS

237 THYLACOSMILUS

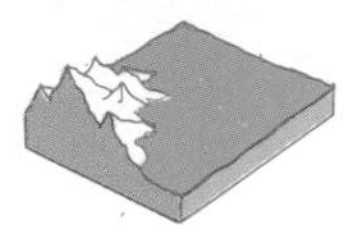

Classification Phylum Chordata, Class Mammalia, Order Marsupialia, Family Borhyaenidae.
Description A marsupial found solely in the South American Pliocene. Of the same dimensions as a tiger, it possessed a very narrow cranium, from which there emerged two highly developed upper teeth (canines) of triangular section and shaped like curved daggers. The molars, both upper and lower, were reduced in size and small in number, as were the premolars. The lower jaw folded down in its anterior section, as if to form an internal protection against the long canines. The skeleton is known only in part: the exact morphology of the hind limbs is unknown. The limbs, however, were massive and the animal had a short tail.
Stratigraphic position and geographical distribution The genus *Thylacosmilus* has been discovered in 4-million-year-old Pliocene sediments in South America. The species *T. atrox* comes from Argentina (the photograph shows a detail of its cranium).
Note *Thylacosmilus* represents a carnivorous form of marsupial. It was a good hunter. Its extinction was probably provoked by the arrival, through Panama, of placental carnivores into the continent of South America.

---

238 DIPROTODON

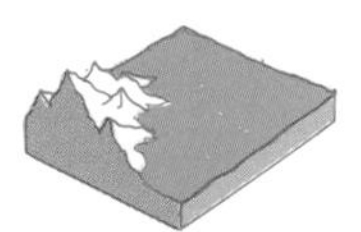

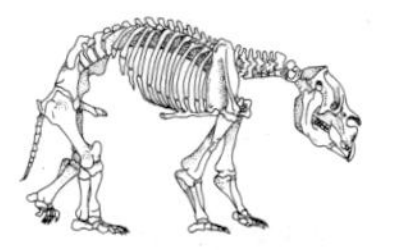

Classification Phylum Chordata, Class Mammalia, Order Marsupialia, Family Diprotodontidae.
Description A large, herbivorous animal characterized by a big cranium equipped with highly developed, forward facing upper and lower incisors; these teeth recall those of a rodent. The upper incisors were curved, the lower ones straight. The cranium was elongate and characterized by the absence of canines. The animal possessed molars, of very simple form, each with two transversal elevations separated by a broad, deep and sharply outlined depression. The front and rear feet, massively built and almost identical in size, probably bore five digits.
Stratigraphic position and geographical distribution The genus *Diprotodon* is known in fossil form from Pleistocene strata in Australia. The example in the photograph (a detail showing the skull) is approximately 1 million years old and comes from these strata.
Note *Diprotodon* is known from fragments only. Australian deposits have yielded numerous skulls, some more than 1 meter (39 in) long, but skeletons are very rare and for the most part incomplete; it is not known, for example, exactly how many digits there were on either its hand or its foot. The representatives of this family vary in size from large to very large.

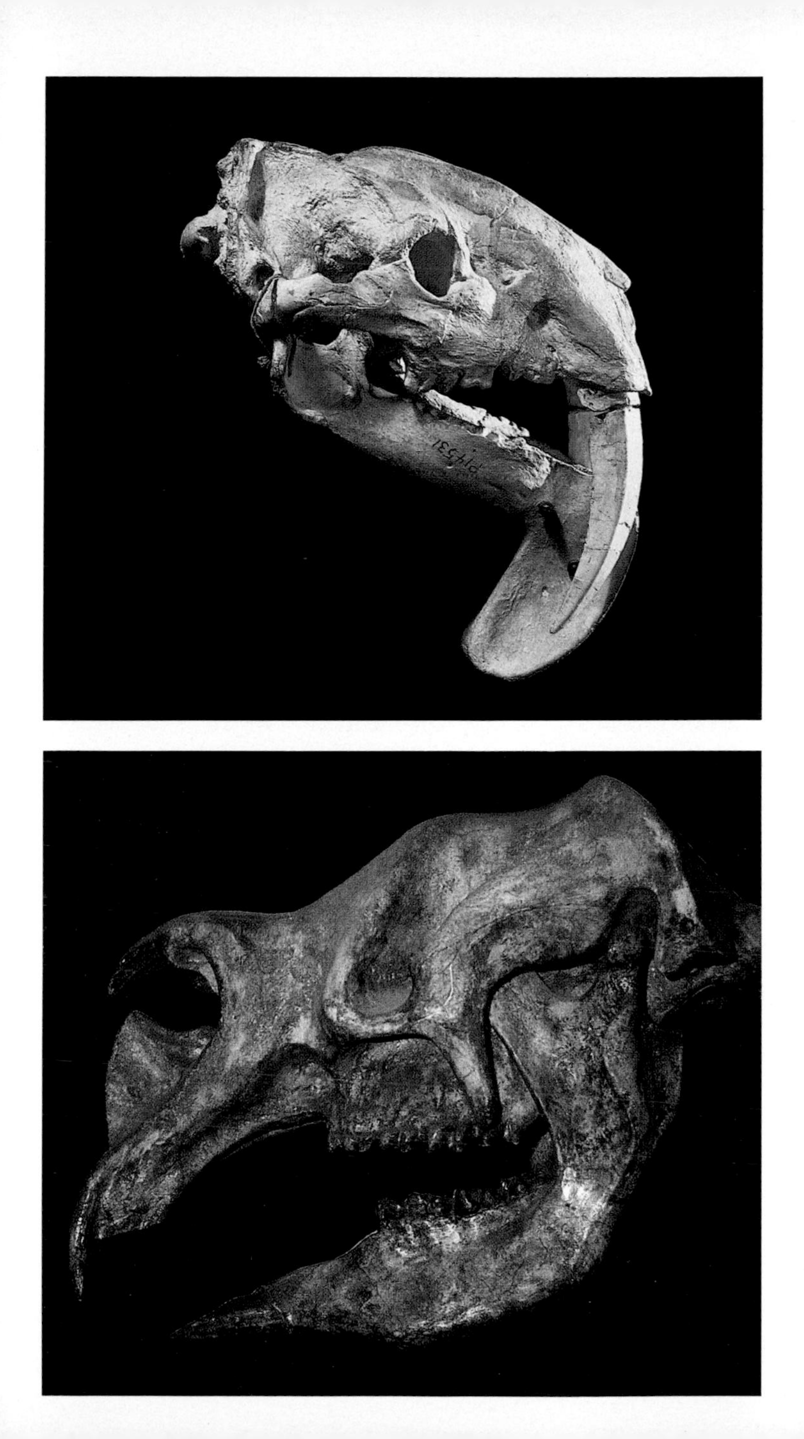

239 CROCUTA

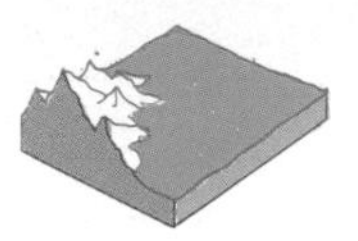

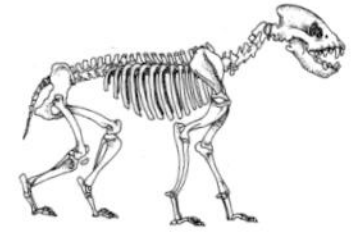

Classification Phylum Chordata, Class Mammalia, Order Carnivoria, Family Hyaenidae.
Description A carnivore of medium size, characterized by a slightly elongate cranium. The dentition includes three incisors, one canine, three premolars, plus a fourth, more developed one: the very long cutting carnassial; there is only one small molar. The body is of agile build, typical of a feline.
Stratigraphic position and geographical distribution The genus is first known from the Upper Miocene (10 million years ago) of Asia, but does not appear until the Lower Pliocene (5 million years ago) in Europe and the Pleistocene in Africa. The detail in the photograph (a lower jaw) belongs to the species *C. crocuta spelaea* and was discovered in caves near Lecce (Italy).
Note *C. crocuta spelaea* is found in great abundance in Pleistocene deposits in English, Italian, Belgian and Spanish caves. It normally lived in groups.

---

240 SMILODON

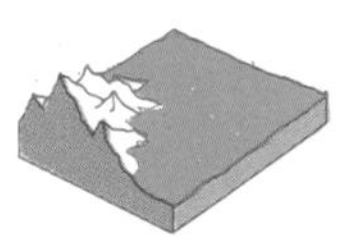

Classification Phylum Chordata, Class Mammalia, Order Carnivora, Family Felidae.
Description The cranium of this powerful carnivore was massive, short and rounded. The dentition was composed of three conical incisors (the numbers apply to each jaw), a highly developed upper canine, vast in size, sharp and shaped like a laterally compressed, backward curving dagger. At the top there were two premolars, one of which, the carnassial, was highly developed, elongate and sharp. Posteriorly there was a single small molar. The skeleton was robust and compact, typically feline in shape. The size of this animal was very near that of a large modern tiger. The feet had four digits, the hands five; the nails were retractable.
Stratigraphic position and geographical distribution The genus *Smilodon* has been discovered in sediments dating from the Upper Pleistocene (1 million years ago) in the U.S. and South America. The species *S. californicus* has been found in the Rancho La Brea tar pits in California, from where the example in the photograph originates.
Note *Smilodon*, commonly known as the saber-toothed tiger, was an active feline capable of hunting down even the great proboscideans of the Pleistocene. Complete examples of this animal have been discovered in the Rancho La Brea tar pits, generally trapped together with its prey.

241 FELIS

Classification Phylum Chordata, Class Mammalia, Order Carnivora, Family Felidae.
Description An active carnivore with a cranium containing three incisors, a strong, conical canine, two premolars—in some cases three—and a single molar. The fourth premolar, the carnassial, is elongate and characterized by a well-developed internal tubercle. The molar is of modest dimensions. Between the two lower premolars and the canines there is a diastema. The animal also has a tail.
Stratigraphic position and geographical distribution The genus *Felis* is first known from the Lower Pliocene (5 million years ago) on the continents of Europe and Asia. Only in the Pleistocene did it spread through North America, South America and Africa. The detail in the photograph belongs to the species *F. leo spelaea* and comes from the Pleistocene deposits at Lot (France).
Note *F. leo spelaea*, sometimes known as the cave lion, occurs in large numbers in European caves. The domestic cat also belongs to the genus *Felis*.

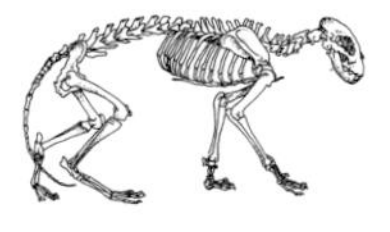

---

242 URSUS

Classification Phylum Chordata, Class Mammalia, Order Carnivora, Family Ursidae.
Description A large, omnivorous carnivore (commonly known as a bear), with an elongate cranium characterized by a very marked sagittal crest. The dentition is composed of premolars that are either very small or absent; the fourth premolar, the carnassial, possesses an anterior conical tubercle, while other molars are broad and robust. The animal has incisors and also conically shaped canines, which are long and sturdy. The limbs are powerful, with long femurs and humeri, while the feet and hands both possess five digits, a feature typical of a plantigrade animal. The tail is short and there is a penis bone.
Stratigraphic position and geographical distribution The genus *Ursus* appears during the Lower Pliocene in Europe (5 million years ago) and during the Pleistocene it reached North Africa, Asia and North America. It still lives in Europe and North America. The cranium in the photograph, which belongs to the species *U. spelaeus*, comes from Pleistocene caves near Cuneo (Italy).
Note The species *U. spelaeus* was very widespread in Europe during the Pleistocene. It lived in caves in the sides of mountains, and it has now been found in fossil form there. *U. spelaeus* is one of the commonest fossil carnivores in Europe; its size was about the same as the modern American grizzly bear.

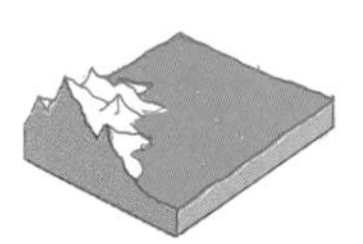

243 UINTATHERIUM

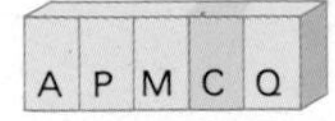

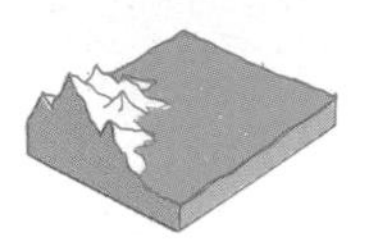

Classification Phylum Chordata, Class Mammalia, Order Amblypoda, Family Uintatheriidae.
Description A large terrestrial mammal of unusual appearance. It possessed a cranium containing three large pairs of bony apophyses situated above the parietal bone, the nasal bone and the maxillary bone, respectively. The brain of this animal was extremely small in size. The unusual dentition lacked upper incisors: these were replaced by enormous canines that pointed downwards and extended over the lower jaw. The incisors and lower canines were small. The limbs were highly developed to cope with the weight of this great animal, whose feet and hands had five, well-developed digits.
Stratigraphic position and geographical distribution The genus *Uintatherium* is known in fossil form from Middle and Upper Eocene sediments (50–40 million years ago) in the states of Wyoming, South Dakota and Nebraska. The species *U. mirabile*, whose cranium is shown in the photograph, comes from Wyoming.
Note *Uintatherium* has been discovered in great abundance in the Bad Lands, which are composed of sediments of fluvio-lacustral origins occurring in the southwestern part of South Dakota and the northwestern part of Nebraska. It was a herbivorous animal.

---

244 ANANCUS

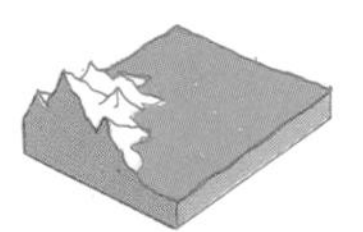

Classification Phylum Chordata, Class Mammalia, Order Proboscidea, Family Gomphotheriidae.
Description A proboscidean of large dimensions, characterized by a cranium similar to that of *Elephas*, but slightly higher and with a lower and more elongate mandible. The upper incisors are inserted in deep alveoli and are long and straight, only very rarely curving. The upper molars are large and of elongated quadrilateral form, with three or four, sometimes five or six, transversal ridges separated by deep winding depressions. The lower molars, by contrast, are smaller, with no gaps between one crest and the next; this flattening is probably the result of wear. There are no lower incisors.
Stratigraphic position and geographical distribution The genus *Anancus* is known from the beginning of the Lower Pliocene (5 million years ago) in Europe and Asia; during the Pleistocene it reached Africa. The photograph shows a molar of *A. arvernensis* from the Pliocene at Paderno d'Adda in Como (Italy).
Note The best preserved remains of this animal are undoubtedly the ones discovered near Montecarlo (Upper Valdarno in Italy), which have been attributed to the Lower Villefranchian (2½ million years ago).

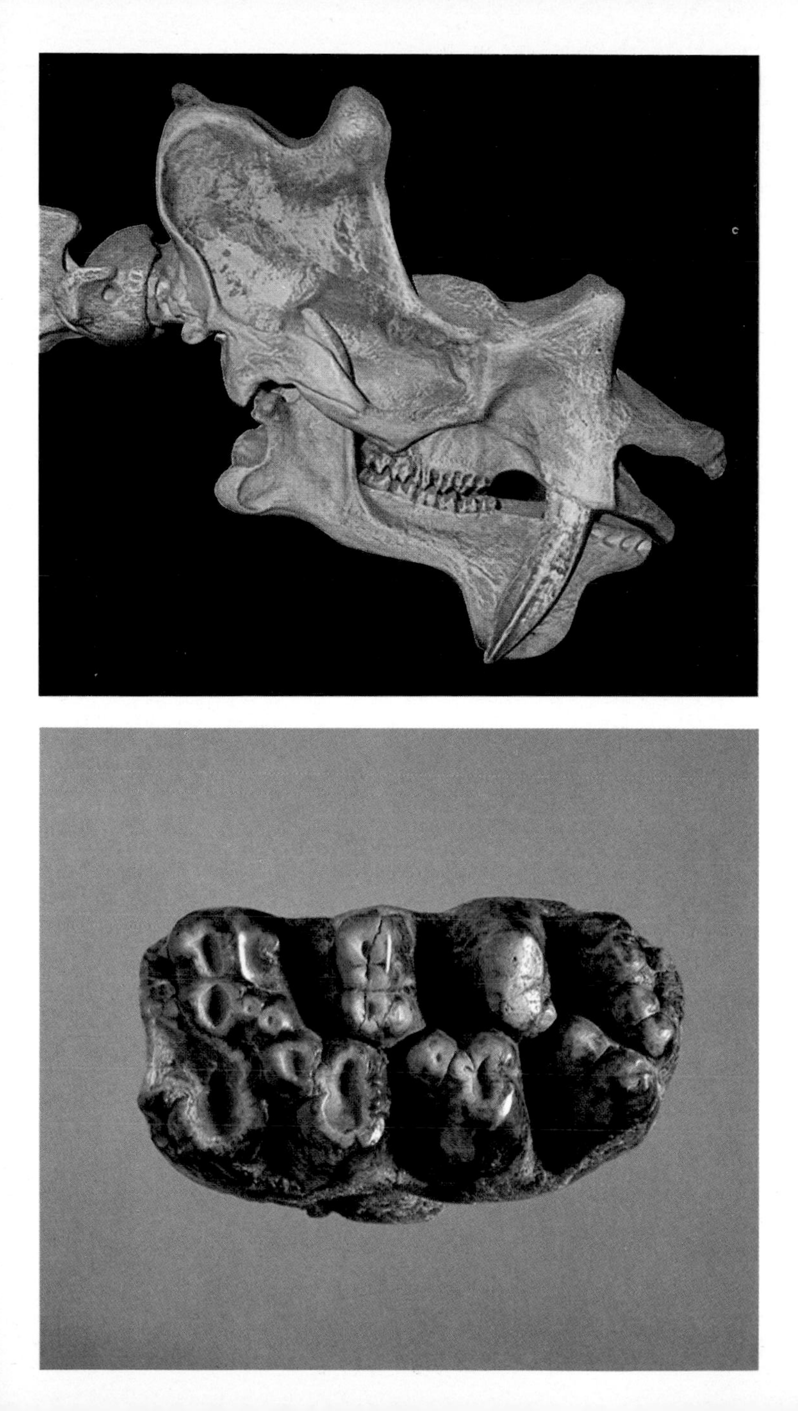
c

245 GOMPHOTHERIUM

Classification Phylum Chordata, Class Mammalia, Order Proboscidea, Family Gomphotheriidae.
Description A very archaic proboscidean characterized by a strongly elongate, protruding mandible that looked like a stout shovel. Its mandible bore very strong incisors, broad and flattened, which also projected forward and were probably used by the animal to uproot the plants on which it fed. The body was large and possessed a short tail, a feature common to all proboscideans, whereas the actual proboscis was negligible. The animal had six teeth above and five below.
Stratigraphic position and geographical distribution The genus *Gomphotherium* is known from the Lower Miocene (22½ million years ago) and has been discovered in Europe, Asia and Africa. During the Upper Miocene it reached North America. It vanished from Europe in the Lower Pliocene (5 million years ago), but on the other continents it did not become extinct until the Pleistocene. The example in the photograph comes from the German Miocene.
Note Although *Gomphotherium* originated on the African continent, it also became widespread on other continents, including North America, where large numbers of fossil skeletons have been discovered. The genus *Stegomastodon*, a member of this family, also reached South America.

---

246 ELEPHAS

Classification Phylum Chordata, Class Mammalia, Order Proboscidea, Family Elephantidae.
Description A proboscidean with a convex cranium characterized by the absence of premolars and the presence of three highly developed molars. The upper incisors have assumed colossal dimensions and become tusks. The molars reveal laminae running transversally to the length of the actual tooth, which is massive and rectangular in shape; the number of these laminae and their progress are features that vary with the animal's age and species. *Elephas* is of gigantic build. The limbs have a humerus much more developed than the femur, with broad feet and hands to carry the weight of the animal. There is a small tail, and the proboscis is functional.
Stratigraphic position and geographical distribution The genus *Elephas* is known from the Lower Pleistocene (1.8 million years ago). Widespread in Europe during the Pleistocene, the genus is now found in Asia. The examples in the photograph, belonging to the species *E. falconeri*, come from the Sicilian Pleistocene and are a little more than 1 meter (39 in) high.
Note The species *E. falconeri* derives from *E. antiquus*, the largest proboscidean ever to have lived in the interglacial periods of the Pleistocene. *E. falconeri* represents a dwarf form that inhabited the islands of the Mediterranean.

247 DEINOTHERIUM

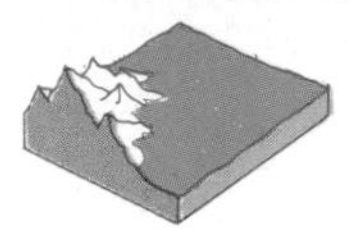

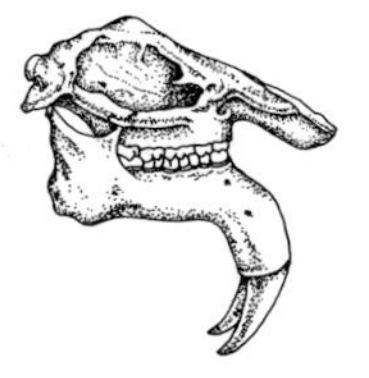

Classification Phylum Chordata, Class Mammalia, Order Proboscidea, Family Deinotheriidae.
Description A unique proboscidean, characterized by a cranium bereft of upper canines and incisors. The upper molars are of quadrangular shape and reveal two ridges, which run subparallel to the longitudinal axis of the tooth and are separated by a deep incision. The lower molars are of completely analogous form. The symphysis of the lower jaw bends downwards, almost at a right angle, with a colossal, pointed, backward curving tusk at each side. The overall appearance and size of this proboscidean does not differ greatly from that of the modern African elephant.
Stratigraphic position and geographical distribution The genus *Deinotherium* is first known in the Lower Miocene (22½ million years ago) in Europe, where it survived until the Upper Pliocene (2 million years ago). It was widespread in Africa during the Lower Miocene and up to the Pleistocene, whereas in Asia it appeared during the Upper Miocene and became extinct in the Upper Pliocene. The photograph shows a molar of *D. giganteum* from the German Upper Miocene at Westhofen.
Note *Deinotherium* probably used its tusks as digging implements; it was herbivorous, feeding on roots and leaves.

---

248 ARSINOITHERIUM

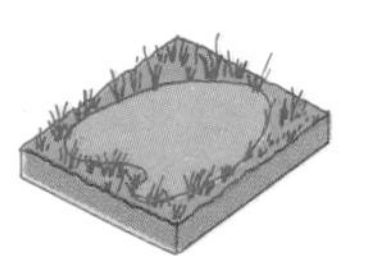

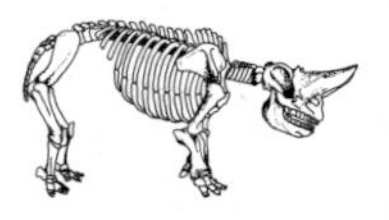

Classification Phylum Chordata, Class Mammalia, Order Embrithopoda, Family Arsinoitheriidae.
Description A unique subungulate of the same size as a rhinoceros. It possessed very solid limbs, a tail and five digits on each foot. It was characterized by the presence of two massive horns on the nasal bones; these protuberances were projected forward and fused together at their base. The dentition was complete.
Stratigraphic position and geographical distribution The genus *Arsinoitherium* has been found in sediments of the Lower Oligocene (35 million years ago) in North Africa. The photograph shows a detail of the cranium of an example found at Fayum (Egypt).
Note This strange animal has been discovered solely in Egypt; its ancestors and descendants are both unknown. It is the only genus belonging to this family, which is why it has been placed in an order of its own, and the genus is subdivided into two species. It appears that this animal led an amphibious life, like the hippopotamus, and was fairly clumsy on land.

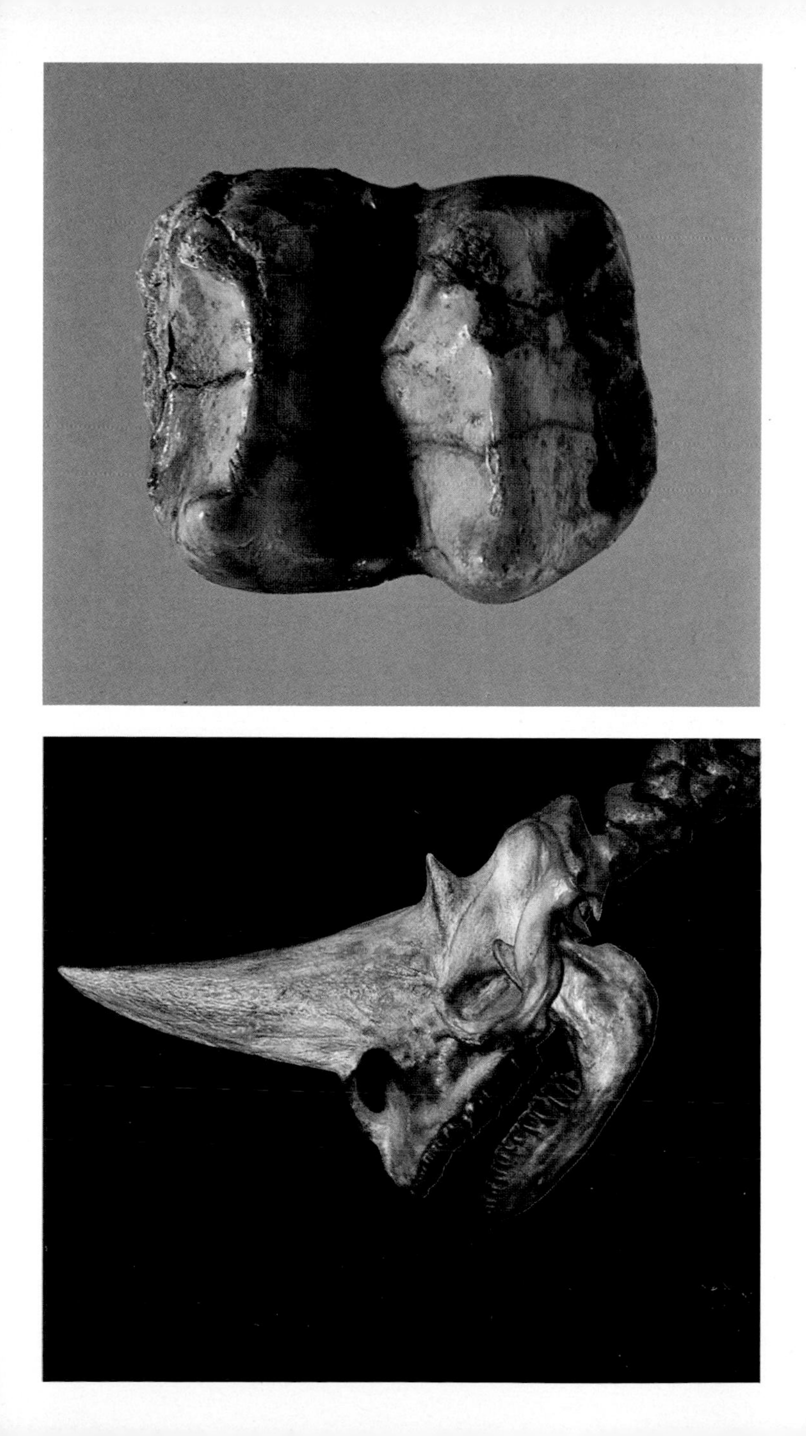

249 PALAEOTHERIUM

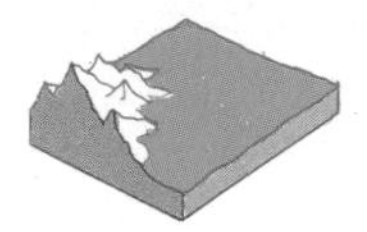

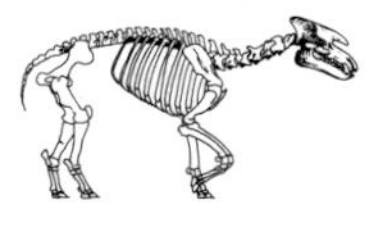

Classification Phylum Chordata, Class Mammalia, Order Perissodactyla, Family Palaeotheriidae.
Description *Palaeotherium* was a perissodactyl of very similar appearance to a tapir; it possessed a high elongate cranium with triangular nasal bones. The eye sockets were very small and positioned far back on the head. The dentition of each mandible was composed of three molars, four premolars, a canine and three incisors. The forelimbs and hind limbs each bore three digits.
Stratigraphic position and geographical distribution The genus *Palaeotherium* appeared in the Upper Eocene (40 million years ago) in Europe and became extinct shortly afterwards. It occurs in France, England, Germany and Switzerland. The jaw fragment in the photograph comes from the layer at Saint-Saturnin-d'Apt (France).
Note *Palaeotherium* was also found in the gypsum beds at Montmartre in Paris during the excavations carried out for the building of that famous district. One complete specimen, with its bones intact anatomically, was discovered and studied by the great paleontologist Georges Cuvier. The species unearthed in Paris was called *P. magnum* because of its great size.

---

250 PLAGIOLOPHUS

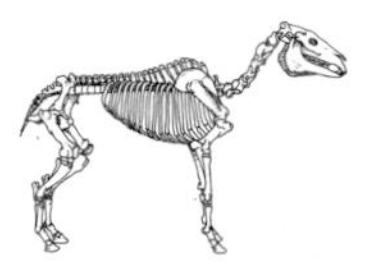

Classification Phylum Chordata, Class Mammalia, Order Perissodactyla, Family Palaeotheriidae.
Description A perissodactyl very similar to a donkey. Its cranium was narrow and elongate, like that of *Palaeotherium*. Its dentition was composed of three fairly well-developed molars, accompanied by three premolars of medium size and a fourth, very small one that was absent in most species. The diastema was fairly well developed. The canines were short and slightly elongated and incisors were also present. All four feet bore three digits.
Stratigraphic position and geographical distribution The genus *Plagiolophus* ranges from the Middle Eocene (50 million years ago) to the Lower Oligocene (35 million years ago) in Europe. The photograph shows a lower jaw of *Plagiolophus* from Gargas (France).
Note The earliest species of this genus were characterized by the presence of a developed first premolar. Complete skeletons of this animal have been unearthed in the gypsum beds of Montmartre in Paris.

251 BRONTOTHERIUM

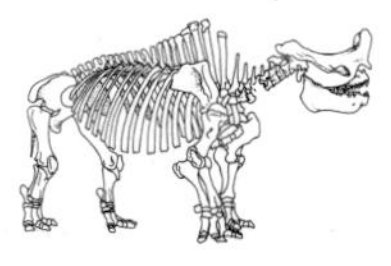

Classification Phylum Chordata, Class Mammalia, Order Perissodactyla, Family Brontotheriidae.
Description A gigantic animal that measured more than 2.5 meters (8 ft) at the withers, with the overall dimensions of an elephant. It possesses a depressed, elongate and very broad cranium with osseous protuberances, particularly on the nasal bones. Because of their shape and length, these protuberances are used both in generic systematics and also as a means of distinguishing the sexes in this animal. The dentition is composed of very small incisors, which are sometimes absent, and molars located immediately behind the canines and increasing in size as they near the back of the mouth. The limbs are very sturdy and bear short, strong phalanges. There are four digits on the forelimbs and five on the hind ones. The animal possesses a short tail.
Stratigraphic position and geographical distribution The genus occurs in Lower Oligocene strata (35 million years ago) in North America. The fragment in the photograph (part of the dentition) comes from the state of Wyoming.
Note The White River Beds of the states of Nebraska, South Dakota and Colorado have yielded abundant remains of this animal.

---

252 MOROPUS

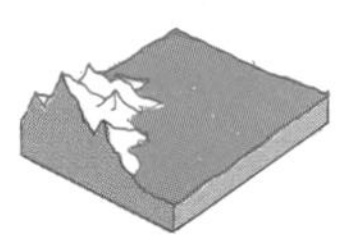

Classification Phylum Chordata, Class Mammalia, Order Perissodactyla, Family Chalicotheriidae.
Description *Moropus* was a mammal (the size of a large horse) that fed on plant and vegetable matter. It was characterized by a fairly well-developed cranium borne on a long mobile neck, with highly specialized molars and premolars. Both the forelimbs and hind limbs had only three digits, each equipped with highly developed claws. The animal was plantigrade. One particular characteristic of this herbivore was its slightly bent hind legs, the forelegs remaining completely straight, which made the standing animal look much lower in the back than the front. Its tail was poorly developed.
Stratigraphic position and geographical distribution The genus *Moropus* appears in the Lower Miocene (22½ million years ago) in North America and was widespread on that continent until the Upper Miocene (10 million years ago), when it vanished without leaving any direct descendants. Some fossil remains have also been found in Asia. The example in the photograph comes from the Miocene in the state of Nebraska.
Note The family to which this gigantic mammal belongs contains grotesque-looking animals, some of whose forms also spread through Europe.

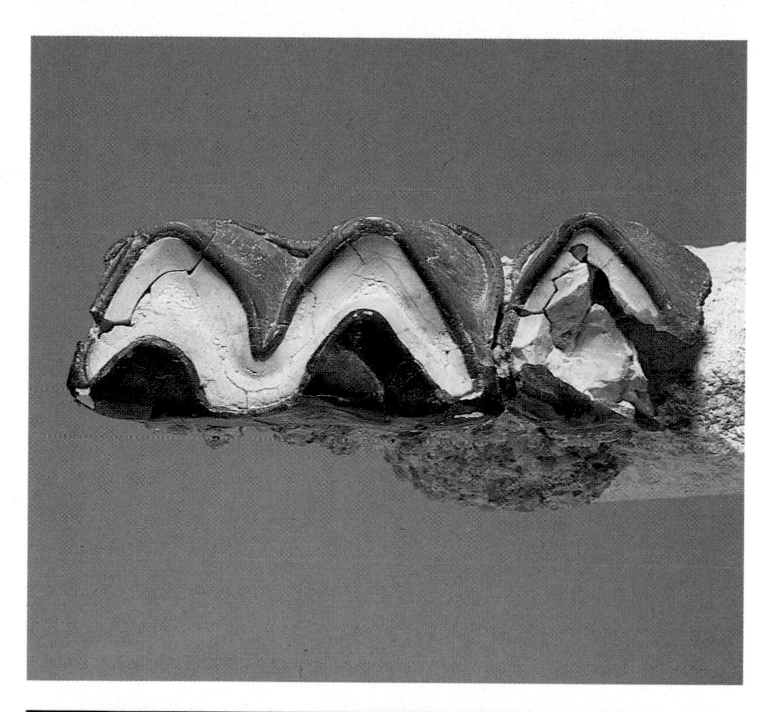

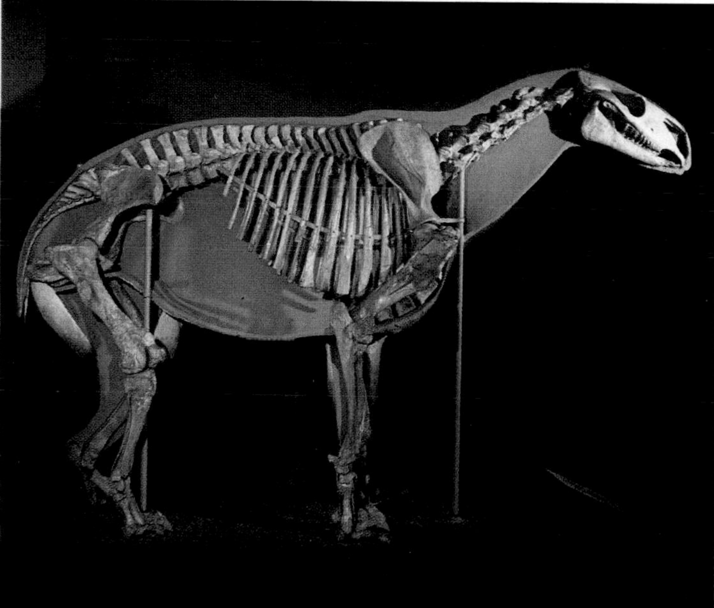

253 HYRACODON

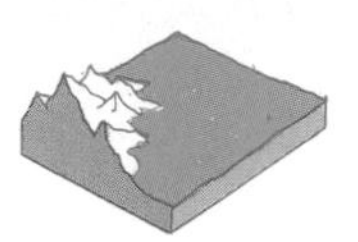

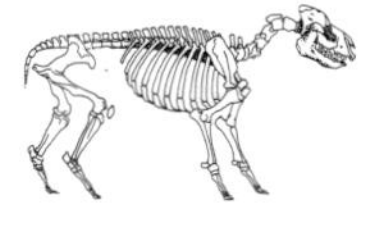

Classification Phylum Chordata, Class Mammalia, Order Perissodactyla, Family Hyracodontidae.
Description *Hyracodon* was a mammal whose general appearance was very similar to that of a small rhinoceros. The cranium had a very long sagittal crest, highly developed nasal bones and no horn. It had a complete dentition, with a full quota of incisors, which were subconical in shape, fairly well-developed canines, four premolars and three molars. The general structure of the spinal column was similar in every respect to that of the modern rhinoceros. The forelimbs and hind limbs bore three laterally compressed digits.
Stratigraphic position and geographical distribution The genus *Hyracodon* appears in the Lower Oligocene (38 million years ago) and became extinct in the Lower Miocene (20 million years ago). Its geographical distribution is restricted solely to the continent of North America. It occurs particularly in the states of Colorado and Nebraska. The mandible fragment in the photograph comes from Nebraska.
Note This primitive rhinoceros was much smaller and more graceful than the modern rhinoceros. It is believed that *Hyracodon*, lacking any means of defense, became a good runner to escape the attacks of its predators.

---

254 ANTHRACOTHERIUM

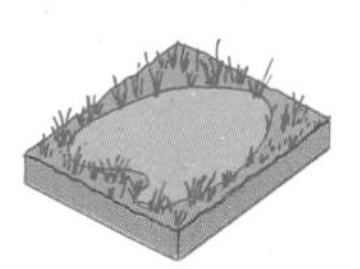

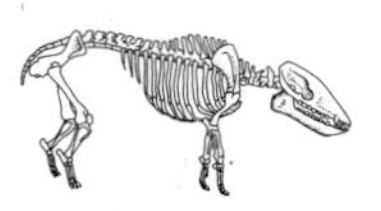

Classification Phylum Chordata, Class Mammalia, Order Artiodactyla, Family Anthracotheriidae.
Description An animal with a body whose general appearance resembles that of a pig. The cranium is low, very elongate and has a parietal crest. The eye sockets are small and placed at the back of the head. The facial section dominates, being larger and longer overall than the rest of the cranium. The dentition is composed of three molars and four premolars on each jaw, a highly developed canine, of conical form; the incisors are paddle-shaped and projecting. The limbs are stout and similar to those of a hippopotamus, though slightly thinner.
Stratigraphic position and geographical distribution The genus *Anthracotherium* has been discovered in strata of the Lower and Medium Eocene (54–43 million years ago) in Europe, occurring in France, Italy, Switzerland and Germany. Some species managed to reach Asia, where they did not become extinct until the Lower Pliocene (5 million years ago). The photograph shows a detail of the dentition of *A. magnum* from Cadibona (Italy).
Note This mammal spent much of its time immersed in the waters of rivers and lakes, a way of life much like that of the modern hippopotamus.

255 BOTHRIODON

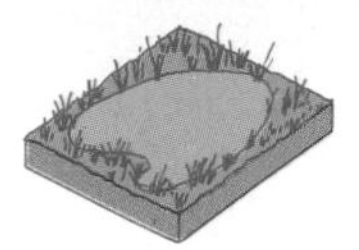

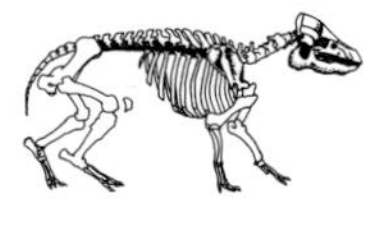

Classification Phylum Chordata, Class Mammalia, Order Artiodactyla, Family Anthracotheriidae.
Description A very primitive artiodactyl, common during the Tertiary, whose form is very similar to that of a pig. It did not exceed 1.5 meters (5 ft) in length. It has short limbs, with four working digits (in the earliest forms there is also a fifth digit). The cranium is low, with a very narrow and elongate snout. The dentition is complete, with normally developed canines and incisors, four premolars, the last three of which are well developed, and three molars of quadrate shape on each jaw. The eye sockets are small and situated towards the back of the head.
Stratigraphic position and geographical distribution The genus *Bothriodon* has been found in Eocene strata (50 million years ago) in Europe and Asia; during the Oligocene it became widespread in Europe and North America. It did not reach Africa until the Miocene. The photograph shows a detail of the dentition of *Bothriodon* from the North American Oligocene.
Note This form is very similar to *Anthracotherium*, from which it is distinguished by certain osteological details. It inhabited the banks of lakes, rivers and marshes.

---

256 HIPPOPOTAMUS

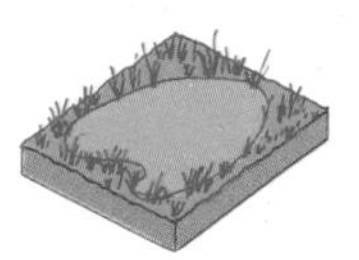

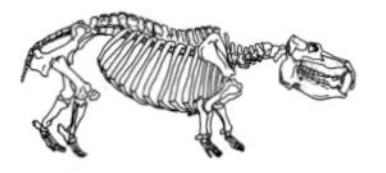

Classification Phylum Chordata, Class Mammalia, Order Artiodactyla, Family Hippopotamidae.
Description An ungulate of large dimensions, with a complete bunodont dentition. The upper and lower molars bear four tubercles, whereas the premolars are much simpler. The upper canines are very thick, short and projecting; the lower canines are of enormous size and form an upward curving arch. The lower incisors are cylindrical, very long and point forward. The massive cranium is low and elongate; it bears small eye sockets placed high on the head. The limbs are short, squat and powerful.
Stratigraphic position and geographical distribution The genus *Hippopotamus* appeared in the Upper Pliocene (3 million years ago) in Asia and Africa; it has vanished from Asia in recent times. During the Pleistocene interglacial periods some forms reached Europe. The photograph shows a detail of the dentition of *Hippopotamus* from the Quaternary at Valdarno (Italy).
Note The sole genus of this living family is restricted to tropical Africa. It is a large herbivore, living an amphibious life, and is a descendant of the Anthracotherian group.

257 MERYCOIDODON

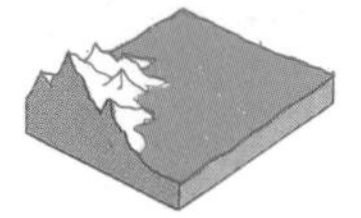

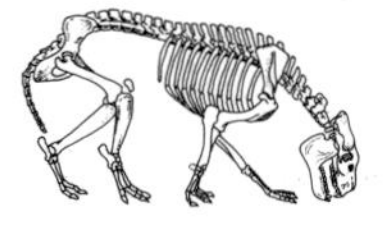

Classification Phylum Chordata, Class Mammalia, Order Artiodactyla, Family Merycoidodontidae.
Description A herbivorous animal of medium size, with a subrectangular, anteriorly truncate cranium. The dentition is complete and includes moderately developed conical canines. This strange ruminant bears five digits on the forelimb and four differentiated digits on the hind limb. The phalanges are short and flattened, much more like those of a carnivore than a herbivore. There are a large number of caudal vertebrae.
Stratigraphic position and geographical distribution The genus *Merycoidodon* is found in strata of the Lower and Middle Oligocene (35–30 million years ago) in the state of Wyoming, the origin of the cranium in the photograph.
Note The dimensions of *Merycoidodon* are similar to those of a peccary. This archaic artiodactyl occurs in abundance in the White River Beds in the states of South Dakota, Nebraska and Wyoming. There are four known species, distinguished mainly by their size: some of them are smaller than an ordinary pig.

---

258 MEGALOCEROS

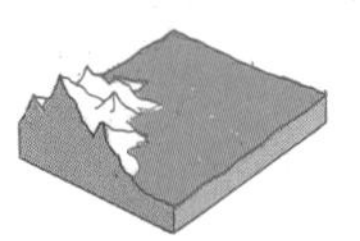

Classification Phylum Chordata, Class Mammalia, Order Artiodactyla, Family Cervidae.
Description A large elk, characterized by an elongate cranium which on its posterior section bears two horn antlers of truly spectacular horizontal dimensions, sometimes measuring more than 3 meters (9¾ ft) across. The dentition lacks incisors, but possesses very strong molars and premolars with a cutting surface typical of ruminants, ideal for grinding down plant and vegetable matter. The large eye sockets are laterally placed in the middle of the cranium. The limbs are highly developed, with feet evolved for jumping.
Stratigraphic position and geographical distribution The genus is typical of the European Pleistocene (2 million years ago) and only recently became extinct. It also became extinct in North Africa. The photograph shows a cranium of *Megaloceros* from alluvial sediments of the Lambro river in Lombardy (Italy); it measures 55 cm (21¾ in).
Note Complete skeletons of this gigantic Quaternary elk have been discovered in peat bogs in Ireland. It was a contemporary of prehistoric man, who probably hunted the animal by driving it into woods, where its large antlers would have prevented it from maneuvering properly.

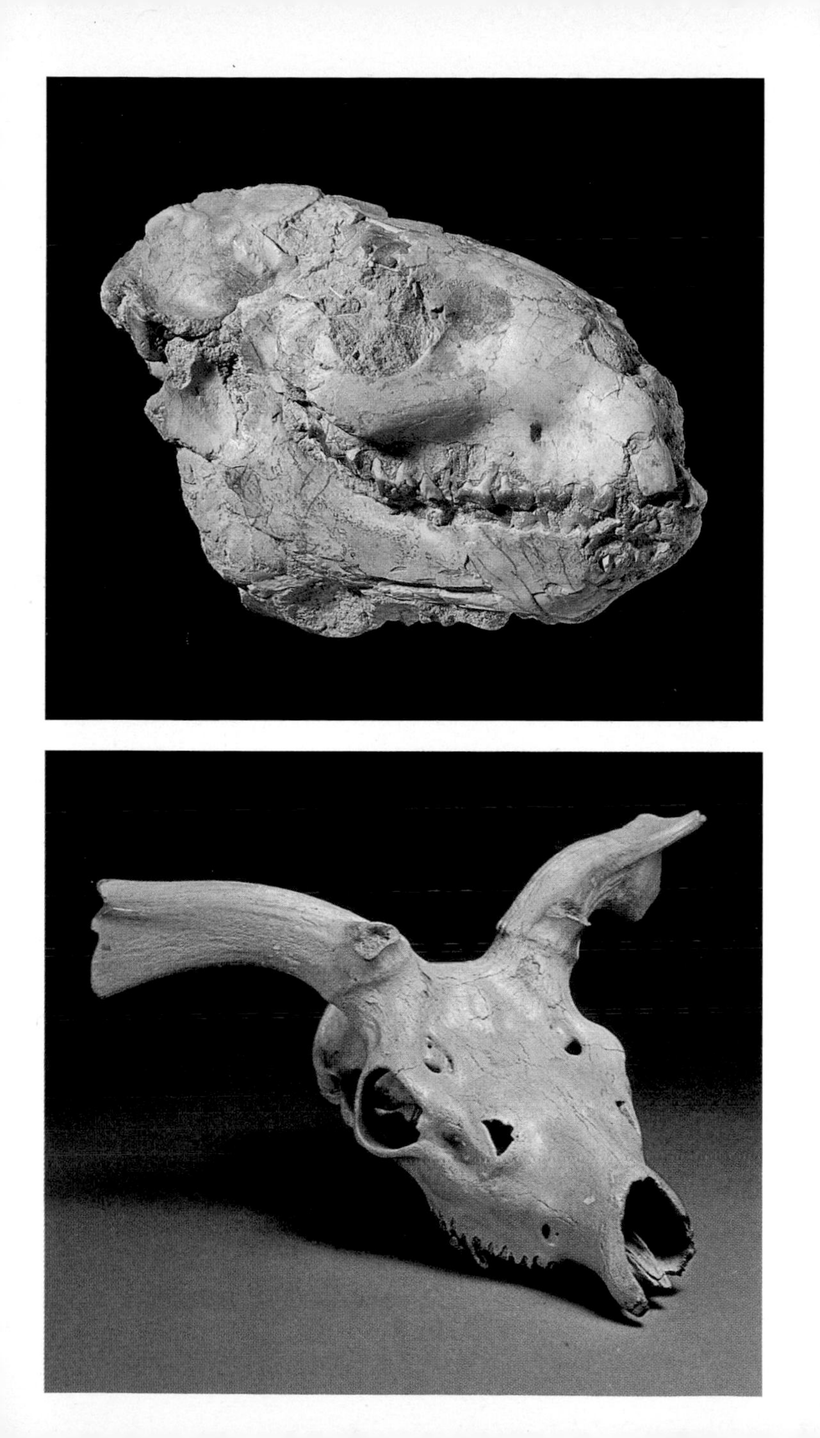

259 BOS

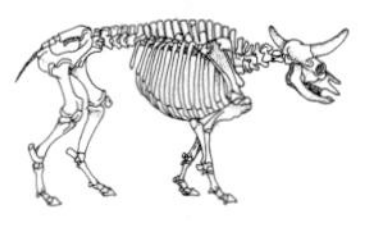

Classification Phylum Chordata, Class Mammalia, Order Artiodactyla, Family Bovidae.
Description A ruminant mammal of large dimensions, with a smooth, low and strongly elongate cranium, whose frontal bone is so broad that it forms the whole of the cranial arch; the parietal bones are very short. The horns, which are cylindrical in shape, are inserted far back on the outer posterior region of the skull. The eye sockets are large and laterally sited at the center of the skull. The dentition consists of molars and premolars, and there is a long diastema. Only the lower jaw possesses incisors.
Stratigraphic position and geographical distribution The genus *Bos* appears in the Pleistocene (1.5 million years ago) and is still extant today; its geographical distribution is restricted to Europe, Asia, Africa and Alaska. The photograph shows the cranium of *Bison priscus* from Pleistocenic deposits at Arena Po (Italy); it measures 1.05 meters (3½ ft).
Note *B. primigenius* is also known as the auroch, which was widespread during the Quaternary and was probably hunted by prehistoric man.

---

260 GLYPTODON

Classification Phylum Chordata, Class Mammalia, Order Edentata, Family Glyptodontidae.
Description A large mammal, which could grow as long as 2.5 meters (8 ft). It had a hemispherical carapace made of rough, osseous plates of considerable thickness. These plates were pentagonal or hexagonal and joined together by a jagged suture. The plates formed an irregular mosaic on the carapace, with pointed plates at the edges. The cranium, also covered in small plates, was very reduced and characterized by the presence of eight very simple molars on each jaw. The limbs were robust, so as to support the great weight of the animal; the ungual phalanges were equipped with strong nails; the forelimbs had four digits, the hind limb five. The animal had a tail, of variable length, equipped with stout spines and mobile plates.
Stratigraphic position and geographical distribution The genus *Glyptodon* is first known from the Upper Pliocene (4 million years ago) and became extinct in the Pleistocene. Its geographical distribution is restricted to the continents of North and South America. The fragment of carapace shown in the photograph comes from the Argentinian pampas.
Note *Glyptodon* is related to the modern armadillo; the armadillo differs solely in that the plates of its carapace are mobile. *Glyptodon* was a herbivore that successfully generated numerous forms in North and South America. It vanished a little more than one million years ago.

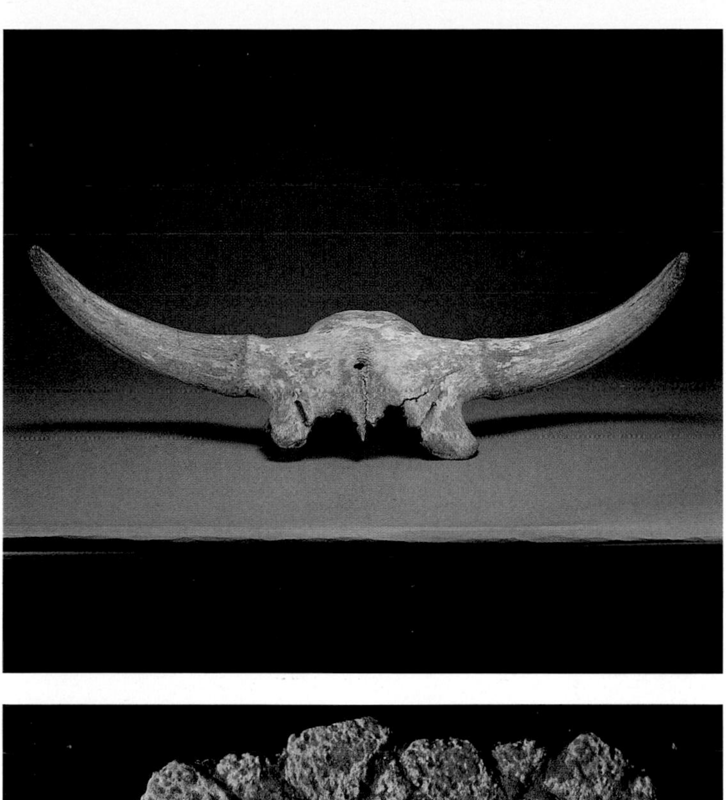

GLOSSARY

Abapical in Gastropoda, from the apex of the shell toward the base.
Abdomen in Crustacea, the posterior region of the body, sited behind the thorax, which also includes the telson; in crustacean Malacostraca, the segments of the abdomen bear limbs. In crabs, the abdomen is folded tightly beneath the thorax.
Aboral in Echinodermata, the surface of the body positioned opposite the surface containing the mouth.
Acicula in Polychaeta, the robust bristles within the parapods.
Adambulacral in Cystoidea, the plate that bears the brachioles.
Adductor muscles in Bivalvia and Brachiopoda, the muscles that allow the valves to close; two adductor muscles are present in articulate Brachiopoda, whereas inarticulate ones possess two pairs.
Ambulacrum in Crinoidea, an elongate area of the oral surface of the body, extending radially from the mouth; in Echinodermata, the structure equipped with brachioles, through which food is diverted toward the mouth, sited on the surface of the exoskeleton.
Ammonitic in Ammonoidea, the type of suture with frilled lobes and saddles.
Anisomyarian in Bivalvia, descriptive of shells with a reduced anterior adductor muscle.
Antennae in Crustacea, consisting of two cephalic appendages that can be uniramous or biramous; they may be very long and composed of numerous small segments, or be reduced or completely absent.
Antennules in Crustacea, the cephalic appendages, mainly filiform, formed of numerous segments; in Malacostraca they are bi- or triramous, but uniramous in other Crustacea.
Anterior in Gastropoda, the direction in which the active animal turns its head.
Anus outlet of the alimentary canal.
Aperture in Gastropoda, the opening surrounded by the margins of the shell.
Apex pointed tip of the shell.
Apical in Gastropoda, toward the apex of the shell along the direction of the axis.
Apical angle in Gastropoda, the angle subtended by two straight lines touching the whorls on opposite sides of the shell in relation to the apex.
Aristotle's lantern masticatory apparatus formed of five teeth, typical of Echinoidea.
Arm in Crinoidea, an elongate structure of the body that protrudes from the radial plates.

Axis, Axial in Gastropoda, imaginary line that passes through the apex of the shell, around which the shell ideally coils; in Trilobita, the median region of the dorsal part of the exoskeleton, bounded longitudinally by a furrow on each side.

Basal in Crinoidea, the circle of plates next to the radial plates bearing the brachioles.
Basal plates in Cystoidea, plates aborally positioned in a circle.
Bifurcate the division into two branches on the ventral region of the costae (ribs) that adorn the shell of Ammonites.
Biplicate fold in Brachiopoda, a fold accompanied by two submedian folds present on the brachial valve and separated by a median furrow (sulcus).
Biramous the subdivision into two branches, one outer (exopodite) and one inner (endopodite), of appendages of Crustacea.
Body cavity in Brachiopoda, the main part of the coelom (body cavity), which is posteriorly sited, bounded by the parts of the body and containing food residues.
Body chamber terminal part of the shell of Ammonites, in which the organism lived.
Brachial valve in Brachiopoda, the valve containing the skeletal supports of the lophophore, with no apertures for the pedicle; usually less developed than the pedicle valve.
Brachioles in Echinodermata, erect structures by means of which food is diverted to the mouth.
Byssus in Bivalvia, a tuft of fibers by means of which the animal can affix itself to foreign bodies.

Callus thickened shell structure in Gastropoda, sited on the parietal region or extending from the inner lip on the base or on the umbilicus.
Carapace in Crustacea, the mineralized structure that covers the head and trunk; in Trilobita, a mineralized tegumen that dorsally covers the body, folding laterally in a ventral direction.
Cardinal tooth hinge tooth of Bivalvia, sited near the umbo.
Cardinal margin in Brachiopoda, the posterior margin, usually curving, of the shell, equivalent to the hinge line.
Cardinal process in Brachiopoda, a narrow area or protuberance on the shell situated at the posterior end of the brachial valve, where the diductor muscles are inserted.
Cardinalia in Brachiopoda, the support for the lophophore and for the attachment of muscles.
Cephalon in Trilobita, the anterior portion of the dorsal exoskeleton, posteriorly articulated with the thorax.

Ceratitic type of suture in Ammonoidea with frilled lobes and rounded saddles.
Chela terminal part of an arthropod limb (particularly in Arachnida and Crustacea), composed of both a mobile section and a fixed section to form a claw.
Chilidial plate in Brachiopoda, a triangular surface covering the tip of the notothyrium, usually convex in appearance.
Cirrus in Crinoidea, an articulate appendage, unramified, which projects from the stem.
Collabrale element of shape corresponding to that of the outer lip of gastropod shells.
Columella filled or empty column in gastropod shells, which surrounds the axis of spiral shells and is formed by the inner walls of the whorls.
Commissure in Brachiopoda, the junction between the margins of the valves.
Compressed a shell whose height is greater than its width.
Conispiral a shell turning in a conically shaped spiral.
Costae elements of ornament on mollusc shells consisting of narrow, elongate ribs of varying shape on the outer surface.
Constrictions in Ammonites, depressions that run around the coils of a shell.
Convolute a shell whose last coil completely envelops the preceding ones.
Crura in Brachiopoda, the two processes that extend from the cardinalia to the lophophore.

Delthyrium in Brachiopoda, a median aperture, triangular or subtriangular, bisecting the ventral cardinal area; from this aperture the pedicle emerges.
Deltoid plates in Blastoidea, plates positioned at the top of the calyx surmounting the radial plates.
Depressed of a shell whose width is greater than its height.
Desmodont hinge in Bivalvia, a hinge bereft of proper teeth, which are replaced by costae and furrows on which the ligament is situated.
Diameter in Gastropoda with conispiral shells, the distance between two planes parallel to the shell's axis.
Diastema in Vertebrata, the area of the mouth in which teeth are absent.
Dichotomy type of ramification widespread in primitive plants, in which the apex of the axis subdivides into two secondary axes.
Diductor muscles in articulate Brachiopoda, the two muscles that serve to open the valves; they are commonly composed of two pairs inserted in the brachial valve.

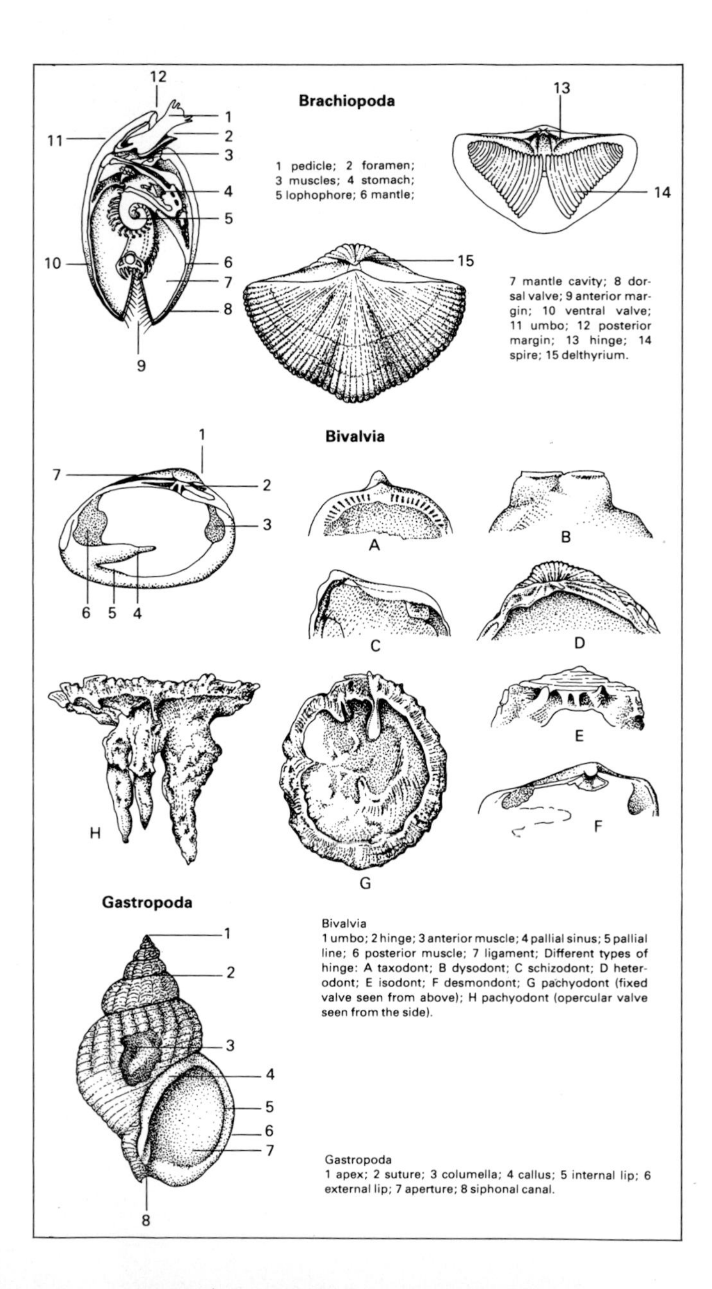

1 pedicle; 2 foramen; 3 muscles; 4 stomach; 5 lophophore; 6 mantle; 7 mantle cavity; 8 dorsal valve; 9 anterior margin; 10 ventral valve; 11 umbo; 12 posterior margin; 13 hinge; 14 spire; 15 delthyrium.

Bivalvia
1 umbo; 2 hinge; 3 anterior muscle; 4 pallial sinus; 5 pallial line; 6 posterior muscle; 7 ligament; Different types of hinge: A taxodont; B dysodont; C schizodont; D heterodont; E isodont; F desmondont; G pachyodont (fixed valve seen from above); H pachyodont (opercular valve seen from the side).

Gastropoda
1 apex; 2 suture; 3 columella; 4 callus; 5 internal lip; 6 external lip; 7 aperture; 8 siphonal canal.

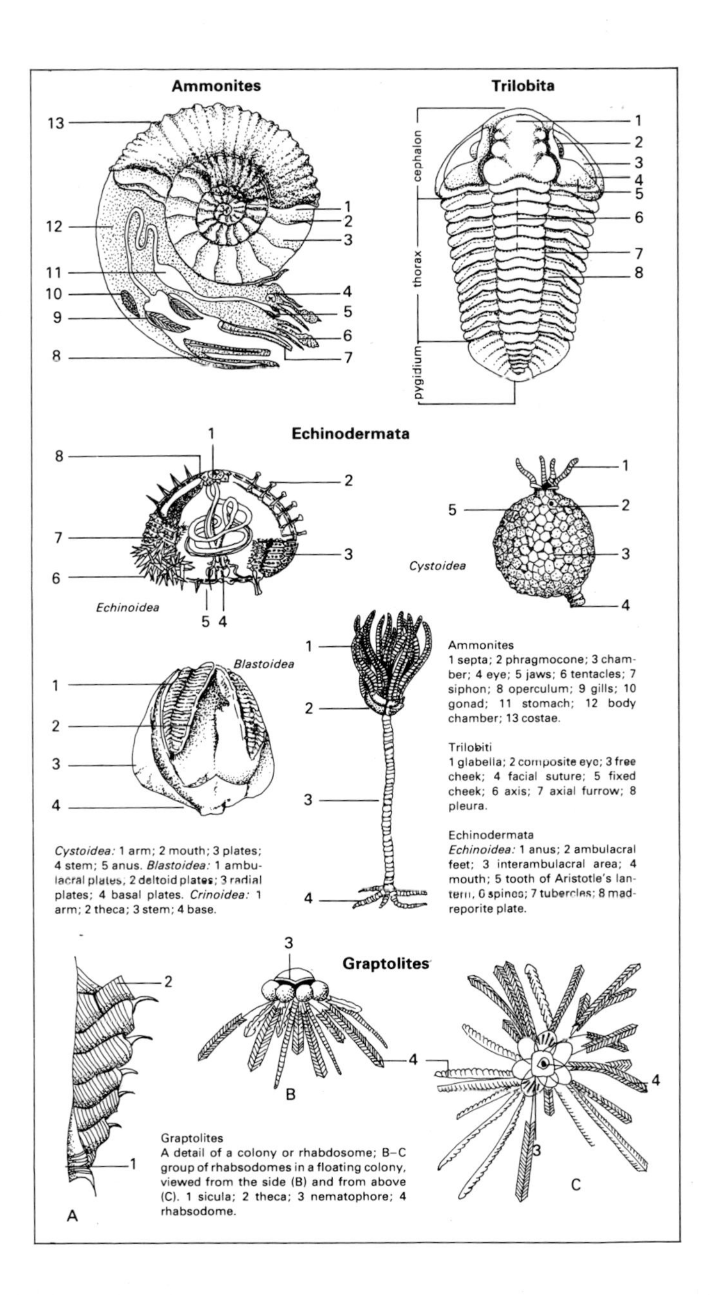

Ammonites
1 septa; 2 phragmocone; 3 chamber; 4 eye; 5 jaws; 6 tentacles; 7 siphon; 8 operculum; 9 gills; 10 gonad; 11 stomach; 12 body chamber; 13 costae.

Trilobiti
1 glabella; 2 composite eye; 3 free cheek; 4 facial suture; 5 fixed cheek; 6 axis; 7 axial furrow; 8 pleura.

Echinodermata
Echinoidea: 1 anus; 2 ambulacral feet; 3 interambulacral area; 4 mouth; 5 tooth of Aristotle's lantern; 6 spines; 7 tubercles; 8 madreporite plate.

Cystoidea: 1 arm; 2 mouth; 3 plates; 4 stem; 5 anus. *Blastoidea:* 1 ambulacral plates; 2 deltoid plates; 3 radial plates; 4 basal plates. *Crinoidea:* 1 arm; 2 theca; 3 stem; 4 base.

Graptolites
A detail of a colony or rhabdosome; B–C group of rhabsodomes in a floating colony, viewed from the side (B) and from above (C). 1 sicula; 2 theca; 3 nematophore; 4 rhabsodome.

Dimyarian bivalve shell with two adductor muscles.
Distal position or direction going away from the axis of the body.
Dorsal exoskeleton the carapace in Trilobita (see **Carapace**).
Dorsum in Ammonites, the innermost part of the shell whorl.
Dysodont hinge in Bivalvia formed of elongate, very modified teeth, similar to folds.

Endopodite in Crustacea, the innermost branch of a biramous appendage.
Equilateral In Bivalvia, when the anterior portion of the valves are of the same dimensions as the posterior one.
Equivalve in Bivalvia, when the two valves are equal both in size and form.
Evolute in Ammonites, a shell whose whorls overlap only a little or not at all.
Exopodite in Crustacea, the outermost branch of a biramous appendage.
Exoskeleton mineralized tegumen covering a major part of the body and appendages of Arthropoda.

Flank in Ammonites, the lateral region of the shell.

Genal spines in Trilobita, retroflex, pointed projections situated at the posterior lateral extremities of the head.
Glabella in Trilobita, the axial portion of the head, which is also convex.
Goniatitic in Ammonites, a type of suture with unfrilled saddles and lobes (except for the ventral lobe).
Growth line in Mollusca, a slightly raised line on the outer surface of the shell, corresponding to the shell's margins at earlier stages of growth.

Head shield in some Arthropoda, the mineralized structure covering the head region.
Height In shells of Gastropoda, the distance between two planes perpendicular to the axis and touching the extremities of the shell.
Heterodont hinge in Bivalvia formed of two or three teeth placed near the umbo (hinge teeth) and accompanied by lateral teeth.
Hypostoma in Trilobita, a plate that covers the mouth or is placed in front of it on the ventral surface.

Inequilateral in Bivalvia, valves whose anterior portion differs in size from the posterior portion.
Inequivalve in Bivalvia, one valve being greater than the other.
Integripalliate in Bivalvia, a pallial line with no pallial sinus.
Interambulacrum in Echinodermata, the area lying between two ambulacra.
Involute in Ammonites, a shell with extensively overlapping whorls; in Gastropoda, a shell whose last whorl covers the preceding ones, which are nonetheless still visible in the umbilicus.
Isodont hinge in Bivalvia, formed of two teeth on each valve, of identical shape and size.
Isomyarian In bivalve shells, when two adductor muscles are of more or less identical size.
Isopygous in Trilobita, indicating that the pygidium is of the same size as the cephalon.

Keel raised structure running the length of the venter on many Ammonite shells.

Lateral plates plates situated on the sides of the theca (in Cystoidea, for example).
Lateral tooth hinge tooth in Bivalvia, placed beside the cardinal tooth.
Leaf lateral expansion of the stem typical of higher plants.
Leaf scar mark left on the tissues of a plant where a leaf has been detached.
Ligament resilient, horny structure which in Bivalvia unites the two valves of the shell and pushes them open; it may be internal or external.
Lobe element of the suture of Ammonites.
Loop in Brachiopoda, the lophophore support that projects forward from the crura.
Lophophore in Brachiopoda, the feeding organ which displays filamentary appendages arranged symmetrically in front of the mouth.
Lunule depression in Bivalvia placed in front of the umbo.

Macropygous in Trilobita, when the exoskeleton has a pygidium larger than the cephalon.
Madreporite porous plate situated at the apex of the exoskeleton of Echinoidea, which allows water to flow into the body.
Mantle in Mollusca, the covering that incloses the soft parts of the animal and secretes the shell.

Mantle canal in Brachiopoda, the supple tubiform expansions of the body cavity within the mantle.
Mantle cavity in Brachiopoda, the anterior space between the valves containing the lophophore.
Maxillipedes in Crustacea, the anterior thoracic appendages, modified to form part of the mouth apparatus.
Micropygous in Trilobita, indicating that the exoskeleton has a pygidium smaller than the cephalon.
Monomyarian used to describe the shell of a bivalve which possesses only a posterior muscle scar.
Multivincular type of ligament found in Bivalvia.
Muscle scars imprints left by the insertion of adductor muscles within the shell of Bivalvia.

Nacre internal structure of a shell, consisting of layers of aragonite running parallel to its inner surface and displaying a characteristic iridescence.
Nematophore in Graptolites, the hollow hemispherical capsule that enables the animal to float.
Neuropodium in Polychaeta, the ventral branch of a parapodium.
Notopodium in Polychaeta, the dorsal branch of a parapodium.
Notothyrium in Brachiopoda, a subtriangular aperture bisecting the dorsal cardinal area or pseudointerarea.

Operculum in Gastropoda, a horny or calcareous structure borne by the foot, which serves to close the aperture.
Opisthogyrous in Bivalvia, when the umbo is backward facing.
Oral in Echinodermata, the surface of the body containing the mouth; in Crinoidea, the five plates arranged in a circle around the mouth.
Ossicle each of the calcareous elements that make up the skeleton of Crinoidea.

Pachyodont hinge in Bivalvia, a derivative of the heterodont hinge, with three highly developed teeth.
Pallial line in Bivalvia, a line on the inner surface of the shell near the margin marking the attachment of the marginal muscles of the mantle.
Pallial sinus indentation in the pallial line, caused by the retractor muscles of the siphon.
Palp in Polychaeta, a sensory or feeding structure on the head.
Parapodium in Polychaeta, a projection of the body positioned on the segments and bearing the bristles.

Pedicle valve in Brachiopoda, the valve from which the pedicle emerges; it is usually more developed than the brachial valve.
Pedicle in Brachiopoda, the structure that emerges from the pedicle valve and is used by the animal to affix itself to the substratum.
Pereiopod in Crustacea, a thoracic appendage with a locomotory function.
Periostracum thin horny layer covering the shells of Mollusca.
Periproct in Echinodermata, the anal aperture in the skeleton.
Phragmocone in Cephalopoda, the part of the shell subdivided by septa.
Planospiral spiral coiling in mollusc shells, typical of Ammonites, in which the whorls lie on the same transversal plane.
Pleopod appendage found on the first five abdominal segments of crustacean Malacostraca (on the first six in Phyllocaridae), often adapted to swimming.
Pleura in Trilobita, the portion of the thoracic segments of the thorax or pygidium, situated laterally to the axis.
Pleural furrows in Trilobita, the furrows on the pleurae.
Posterior in Gastropoda, the opposite direction to the one to which the animal turns its head while active.
Prosogyrous in Bivalvia, when the umbo is forward facing.
Proximal portion or direction turning toward the axis of the body.
Protoconch in Gastropoda, the apical whorl of the shell.
Pseudointerarea in Brachiopoda, the posterior section of the shell.
Pygidium in Trilobita, the posterior portion of the carapace, articulated anteriorly with the thorax.

Radial plates plates that form the side walls of the calyx of Blastoidea and are incised by the ambulacral areas.
Rhabdosome a group of Graptolite structures forming a colony.
Rhizoid elongate, hairlike structure, which in many primitive plants takes the place of true roots.
Rhizome in plants, an underground stem, often horizontal, from which roots and adventitious stems depart.
Root one of the main organs of higher plants, acting as a means of anchorage to the substratum and also of obtaining nutriments.
Rostrum elongate structure projecting in front of the shells of Crustacea, placed in a median position.

Saddle element of the suture of Ammonites, which faces the aperture of the shell.
Schizodont hinge in Bivalvia, formed of a triangular tooth

accompanied, on one valve, by two small lateral pits and, on the other, by two lateral teeth and a central pit.
Sculpture in Gastropoda, the raised ornament on the surface of the shell.
Seed organ that in Gymnospermopsida and Angiospermopsida possesses the ability to produce a new plant; it is derived from a fertile ovule.
Septum in Cephalopoda, the internal structure that divides the shell into chambers and one of whose traces forms the suture line on the exterior surface.
Setae in Polychaeta, the chitinous secretion of the parapods, whose framework it composes.
Sicula Skeleton of the first component member of a graptolite colony.
Siphonal canal in Gastropoda, a narrow, elongate extension of the shell aperture.
Siphonal canal extension of the anterior part of the margin of the gastropod shell containing the inhalant siphon.
Siphuncle in Cephalopoda, the longitudinal tube that runs the length of the phragmocone and opens into the body chamber; in Bivalvia, the tubiform extension of the mantle, allowing the passage of water.
Somite single element into which the body of Crustacea is subdivided.
Spiral ornament on the shell of Gastropoda, running parallel or subparallel to the suture.
Spire in the shell of Gastropoda, the visible, adapical part of all the whorls apart from the last one.
Sporangium small sac which, in primitive plants and some vascular ones, contains cells or spores able to reproduce the organism.
Spore vegetal cell capable of directly germinating a new plant.
Stem the aerial organ of higher plants, bearing the branches, leaves, flowers and fruit, and linking them to the roots.
Stem in Crinoidea, a columnar structure supporting the theca.
Sternite hard ventral cover on the body segments of Trilobita and Crustacea.
Suture in Cephalopoda, the line made by the insertion of a septum into the wall of a shell; in Crinoidea, the line of contact between ossicles; in Gastropoda, the line along which the whorls of a shell come into contact; in Trilobita, the line of juncture of a fixed articulation.

Taxodont hinge in Bivalvia, formed of numerous, small teeth set at right angles to the margin or obliquely placed and conver-

gent toward the umbo.
Telson terminal or anal segment of arthropods (Trilobita and Crustacea) bereft of articulated appendages.
Tergite hard dorsal cover on the body segments of Trilobita and Crustacea.
Theca part of the skeleton of Crinoidea between the arms and the stem; the complex of plates inclosing the body (in Cystoidea for example).
Thorax in Crustacea, the part of the body situated between the head and abdomen; in Trilobita, the part of the body between the cephalon and pygidium.
Trochospiral a shell with a broad base that does not coil in a plane spiral, whose outline may not be sharp.

Umbilicus in Cephalopoda, the central depression on both sides of the shell and bounded by the umbilical wall; in Gastropoda, the cavity around the axis of the shell (assuming there is no filled columella).
Umbo in Bivalvia, the region of the valve coinciding with the point of maximum curvature of the longitudinal dorsal profile.
Uropodite limb on the sixth abdominal segment of crustacean Eumalacostraca, often fan-shaped.

Varice in Gastropoda, the prominent transversal elevation situated on the surface of the shell.
Venter in Ammonites, the outermost region of the shell covering.

Whorl in Cephalopoda, the complete portion of the shell, measuring 360°.

INDEX

(*The numbers are those of the entries*)

Picture sources

P. Arduini—G. Teruzzi, Milan: 11a, b, c, d, e, 12, 22, 23, 43a. S. Cambiaghi, Monza: 21. H. Ibbeken: 41b. Marka, Milan/G. Heilman, Lititz: 8–9, 44–45. Guido Ruggieri, Mestre: 2. Vittorio Salarolo, Verona: 16–17, 25, 27, 29, 31, 33, 35, 36, 37, 39, 41a, 310, 311. Luciano Spezia, Milan: 11f, 14, 43b, 306. Luciano Vanzo, Verona: 13.

Entries
Courtesy Dept Library Services, American Museum of Natural History, New York: 217, 221, 222, 235. P. Arduini—G. Teruzzi, Milan: 6, 7, 8, 9, 16, 26, 33, 39, 41, 45, 53, 56, 60, 67, 70, 85, 87, 88, 92, 93, 94, 102, 103, 104, 109, 112, 113, 116, 118, 121, 124, 125, 127, 132, 138, 139, 146, 147, 149, 163, 186, 194, 206, 224, 236, 245, 246, 260. British Museum, London: 4, 54, 205, 211, 216, 225, 228, 233, 234. Daniela Carli, Verona: all the fossil drawings. Carnegie Museum of Natural History, Pittsburg: 252. P. Dal Gal, Verona: 2, 208, 210, 227. L. Didoni, Monza: 238, 243, 248. Field Museum of Natural History, Chicago: 237. University of Tübingen, Institute of Paleontology: 207. University of Padua, Museum of Paleontology: 213. University of Zürich, Museum of Paleontology: 226. Hauff Museum, Holzmaden-Teck, Württemberg: 230. National Museum of Natural History, Smithsonian Institution, Washington: 240. Giovanni Pinna, Milan: 3. Luciano Spezia, Milan: 1, 10, 11, 12, 13, 14, 15, 17, 18, 19, 20, 21, 22, 23, 24, 25, 27, 28, 29, 30, 31, 32, 35, 36, 37, 38, 40, 42, 43, 44, 46, 47, 48, 49, 50, 51, 52, 54, 55, 57, 58, 59, 61, 62, 63, 64, 65, 66, 68, 69, 71, 72, 73, 74, 75, 76, 77, 78, 79, 80, 81, 82, 83, 84, 86, 89, 90, 91, 95, 96, 97, 98, 99, 100, 101, 105, 106, 107, 108, 110, 111, 114, 115, 122, 123, 126, 129, 130, 131, 133, 134, 135, 136, 137, 140, 141, 142, 143, 144, 145, 148, 149, 150, 151, 152, 153, 154, 155, 156, 157, 158, 159, 160, 161, 162, 164, 165, 166, 167, 168, 169, 170, 171, 172, 173, 174, 175, 176, 177, 179, 180, 181, 182, 183, 184, 185, 187, 188, 189, 190, 191, 192, 193, 195, 196, 197, 198, 199, 200, 201, 202, 203, 204, 209, 212, 214, 215, 218, 219, 220, 223, 229, 232, 239, 241, 242, 244, 247, 249, 250, 251, 253, 254, 255, 256, 257, 258, 259.